T0265805

Rewire

Rewire

Break the Cycle, Alter Your
Thoughts and Create Lasting
Change

NICOLE VIGNOLA

HarperOne
An Imprint of HarperCollinsPublishers

REWIRE. Copyright © 2024 by Nicolesneuroscience Ltd. All rights reserved. Printed in the United States of America. No part of this book may be used or reproduced in any manner whatsoever without written permission except in the case of brief quotations embodied in critical articles and reviews. For information, address HarperCollins Publishers, 195 Broadway, New York, NY 10007.

HarperCollins books may be purchased for educational, business, or sales promotional use. For information, please email the Special Markets Department at SPsales@harpercollins.com.

First HarperOne edition published 2024

Originally published as *Rewire* in 2024 by Michael Joseph

Designed adapted from the Michael Joseph edition

Library of Congress Cataloging-in-Publication Data has been applied for.

ISBN 978-0-06-334979-7

24 25 26 27 28 LBC 5 4 3 2 1

Contents

Contents

Rewire

Intro

Brain matter, your own internal world and the way that it's been programmed

I was a first-year neuroscience undergraduate student the first time I held a brain. I'll never forget my professor telling me to make sure that I kept moving it around in my hands so that I wouldn't leave my fingerprints on it. My mind was utterly blown.

As my thumbs lay over this brain's temporal lobe, I remember thinking about how I was touching every one of this person's memories. The grey matter that we were studying was, in fact, an entire life.

Were they happy?

Were my fingertips touching the exact moment they fell in love?

Did they ever fall in love?

Did they have children?

Was I touching the day they got married? Or perhaps the saddest moment of their life . . . an incident that changed their life forever.

At what point did they decide to donate their body to

science? Maybe they were a scientist too; did they feel the same way when they first held a brain in their hands?

My index and middle fingers were touching their frontal cortex, and I wondered whether they had lived a good life or whether they had been troubled. All these events had taken place before me, and yet I was there, touching them from birth to death. My heart wanted to explode.

For as long as I can remember, I have been trying to answer this question: is it by mistake (by chance) or design? Life, our minds, our beliefs, our habits and behaviours. Is our programming by chance or design? Are our habits and behaviours dictated and chosen by us? Or do we accidentally acquire them through our environment? It wasn't until I completed my degree in neuroscience that I realized the answer is both. We're shaped by the world around us, through external things such as our peers, our circumstances, our religion or culture, and for the most part we should hold on to all those things. We are also shaped by negative thought patterns and narratives we repeat to ourselves, which hold us back from achieving what we want to. We have all believed that we are not good enough at one point in our lives; we have all fixated on something we wish we could change, which prevents us from reaching our fullest potential. Some of these beliefs, habits and behaviours are acquired by chance, and some we have designed for ourselves. But the beauty of the human mind is that no matter how these beliefs were shaped, they can be changed. We can change our programming by our own design. This book is going to explain how our brains are capable of change from a neuroscientific perspective and provide you with a neuro tool-kit that you can apply to your everyday life. The hope we can all hold on to is that we can recreate ourselves at any time and be whoever we want to be, despite what we may have previously believed.

There's a statement I often hear: *I am wired this way.* Scrap that, because by virtue of how the neurons in our brains associate with one another, we can shape our thoughts, habits and behaviours to rewire the subconsciously programmed beliefs we hold about ourselves. The story we tell ourselves about ourselves underpins how we perceive ourselves. This then drives our automatic reactions and behaviours, which, in turn, change how others perceive us.

> *'Our life experience will equal what we have paid attention to, whether by choice or by default'*
> – William James

Weeks after that initial lecture in the neuroanatomy lab, we went back for a dissection class. I remember being tired and struggling to concentrate. My iPhone was running out of battery and it prompted me to switch on 'low-power mode'. And that's when I had another lightbulb moment. The brain is your hardware, and the memories, thoughts, habits and behaviours within it are the software. The software is your mental health and personality, and the hardware is your brain health. Both need to be working optimally to support each other. Aha! *That's* why I was struggling to concentrate!

I was tired and thus the hardware that is my brain wasn't working optimally, and so paying attention and trying to retain information was so difficult. I was essentially operating in low-power mode. My brain was prioritizing the most basic functions so software updates were not important. This made me realize that for someone to upgrade their software and make new connections and retain information, the hardware needs to work well too. That's also why I was struggling to keep up with the pressures of university after having moved to a new city. I was overwhelmed. I wasn't taking care of those fundamental aspects of my health and well-being which directly impacted my brain health. I realized that to positively influence my thought patterns, behaviours and actions, and to retain any information in my lectures, I had to take care of my brain to support my mental health and make new connections.

These connections that each neuron makes, invisible to the naked eye, decide how you're going to execute your day, and they're programmed by the way we live our life. They've also been preprogrammed by our peers and the people before us. This made me realize that if we don't take control of the steering wheel of our life, we're essentially living a life that's been programmed for us. This could be good for us, but for some it could be bad. The good news is that we can upgrade our software. This means we can design new habits and undo unwanted behaviours so that we can reach peak mental well-being and create the best version of ourselves, the version we want to be.

We can make sure our hardware is in good working order by

getting adequate sleep and exercising regularly, something that we will discuss in more detail later. But for now I hope it gives you relief to know that the brain is malleable and it can change.

Perhaps you're wondering how the brain is capable of change. The brain is said to be plastic, derived from the Greek word *plastikos*, which means mouldable; the brain can change and reorganize itself to create new pathways even as we age. Despite the popular belief that we attribute plastic surgery to being plastic like a Barbie doll, this is also why we call it plastic surgery. This is great news for us because it means that our hardware can be improved, and that we can break free from the habits and behaviours that we're so badly trying to shake. I've met a lot of people who feel stuck in their lives, doing a job that they hate because they believe it's their destiny and they only have the skills to do that one thing. Perhaps they trained as an engineer, having been told growing up they had an 'analytical' mind, but they wish they could have gone into something more creative – but *that's not who they are*. They don't realize that neuroplasticity is a thing, so they don't even try to change.

Have you ever thought that you weren't smart enough for something? Or repeated negative things to yourself? And believed that you weren't good enough? Those hypercritical loops of negative self-talk infiltrate your beliefs and are holding you back. You don't even try making a change. So you never apply for college despite the burning desire to learn, because you're afraid you'll fail.

Do you find yourself arguing with your partner for argument's sake because you're 'argumentative'? Do you feel like you create drama wherever you go because that's what your family is like and so you're destined to be the same? Some people even believe that having intrusive thoughts is part of who they are, and that being a negative person is ingrained in their personality. Do you automatically respond with criticism and negativity even when you should feel happy about something positive that's just happened in your life?

These automatic thoughts have been repeated over and over again and become strengthened in the pathways of communication in your brain that shape your way of thinking. And this book is going to show you how to change them.

Shakira was banned from attending choir lessons by her teacher because 'she sounded like a goat' due to her vibrato range. When Beyoncé was a pre-teen, she was in a girl band called Girls Tyme, who competed in *Star Search* in the nineties, the largest talent show on American TV at the time, yet lost. We all know who Beyoncé is now. I think my ninety-year-old Portuguese neighbour even knows who she is. Gisele Bündchen was told at the age of fourteen that her nose was too big and her eyes were too small, so she often got rejected during casting calls. She remembers being told that she would never be on a magazine cover and that it made her feel insecure. Gisele is now one of the highest-paid models in the world with a hugely successful career.

If any of these women had listened to those who tried to hold them back from reaching their full potential, we wouldn't have some of the greatest talents in the music and fashion industries. I want this for you. I want you to be so unapologetic about who you are. I want you to go against the labels you've been given and I want you to have your own story to tell, one that was written by you, not by those who think they know you.

I will equip you with the knowledge and understanding of science to enable you to make long-lasting change in your life. *Rewire* is a neuroscientific toolkit with actionable advice that you can implement in your life and use to change for the better.

I have made it my life's mission to take neuroscience out of the lab and academic journals and make it accessible for everyone. Everybody deserves to know about it because it truly is life-changing stuff. Come with me on a journey and learn to control how you react to whatever life throws at you, from regulating your stress response to overcoming self-limiting beliefs. You can be whoever you want to be.

Create yourself.

How neuroplasticity works

Neuroplasticity is the ability of the brain to change and adapt in response to the internal and external factors that shape it. The connections, functions and structures of the neurons reorganize themselves in response to repetitive input. Neurons are responsible for communicating information in our brain, and the more we repeat something, the stronger the pathways of communication become – like a small footpath that turns into a dirt road that eventually becomes a tarmac highway. Neuroscientists even have a fun phrase for this term: neurons that fire together, wire together. So when we continuously repeat thoughts to ourselves like *I am not good enough*, the brain perpetuates this thought, strengthening it, and it becomes deeply ingrained. Without even realizing it, you're repeating this narrative in the background like the soundtrack to your life. When certain actions, responses or behaviours become ingrained and performed almost instinctively, without a conscious and deliberate decision-making process, we call this automaticity. This happens when we repeat something so that it becomes an automatic response with little cognitive effort or awareness.

For a long time, neuroscientists thought that the brain was incapable of change after we reached a certain age, but more recently we've learned that the brain can change in adulthood. It may be harder to do so in our older years but not impossible; it just requires a little more concerted effort. In fact, the molecular structures of the brain, invisible to the naked eye, all suggest that the brain is designed to change, and it is arguably one of the most important aspects of our neurobiology. It allows us to learn new things, unlearn unwanted behaviours and means we can adapt according to our circumstances and new experiences. The brain can quite literally reorganize itself to create new synapses and form new connections. A synapse is a tiny gap that allows neurons to communicate with one another. At the synapse, one neuron sends a message to another. This is where the exchange of information takes place using chemicals called neurotransmitters. The brain can reorganize itself by forming new synapses and adjusting existing ones through neuroplasticity.

There is even emerging research[1] to suggest that neuroplasticity may be a therapeutic intervention for dementia. Ensuring that we stay engaged with life, both physically and mentally, and continuing to learn new things means that our brains will adapt and maintain their structural integrity. Research[2] shows that older people who remain physically active throughout old age have more proteins in the brain that keep the connections between the neurons strong and healthy. This correlates to higher cognitive function and less neurodegeneration. The brain is a remarkable piece of machinery that we can programme and continue to upgrade so that we can live a full life that's governed by autonomy and control. Many people don't realize the potential that they hold and the power of their brain ... You are in control of your life. I don't want to reach eighty and regret letting life pass me by or letting that inner voice continue to criticize me. I don't want to be eighty and regret letting things happen by chance, on a path that I didn't lay out for myself but was instead governed by my internal rumination. I want to be eighty and scream, 'Hell yeah, I did that! That was my life and I am proud of it because I chose it and didn't settle for it.'

Consider this. You are born with a unique set of DNA and, just like a fingerprint is unique to you, you also have a unique map of brain connections that no other person has. You have a distinct set of identities and traits that may have been buried by your brain's programming, informed by your peers and your circumstances growing up. Let's exhume those traits.

How the brain changes

The three-pound universe sitting in your skull consists of 100 billion neurons with trillions and trillions of connections that work together to create your own internal world, one that is influenced by the outside world yet will never be experienced by anyone else. It's a bit wild when you think about the fact that the only thing separating our internal thoughts from our environment and other people is a quarter-inch piece of bone.

The neurons in your brain communicate with each other at special points called synapses. When neurons communicate with one another repeatedly, they get better at communicating and the signals between them grow stronger. It's why you can find yourself going down a very familiar route with your train of thought; it happens almost automatically. When we repeat certain things, the message between these neurons is transmitted more easily and, before you know it, you've reinforced a belief that seems impossible to shake. Plasticity works both ways, though. We can learn and strengthen pathways but we can also weaken them over time. All the behaviours and patterns you're trying to break can absolutely be dismantled. Throughout this book, and particularly in Phase 1, you will learn how this is done.

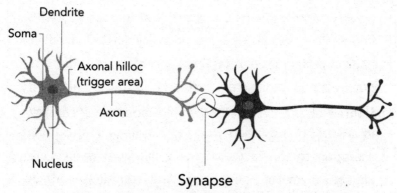

Dendrite

Soma

Axonal hilloc (trigger area)

Axon

Nucleus

Synapse

Neuronal junction where communication happens between neurons. The synapses are strengthened when more communication occurs between them.

A note on neurodiversity

One of the most frequent questions I get from the neurodivergent community on my social channels is whether people who are autistic or who have obsessive-compulsive disorder (OCD) or attention deficit hyperactivity disorder (ADHD) are capable of neuroplasticity. It's important to acknowledge that every single individual, neurodiverse or neurotypical, is completely different, but we are all capable of change. I also want to highlight the challenges that neurodiverse individuals encounter in a world that expects them to change. I aim to bring attention to the fact that neurodivergent people are capable of making new habits and adopting healthier patterns of thinking in a way *that serves them*. This should be if any neurodiverse person wishes to make changes, not to fit into a neurotypical world. While the brains of neurodivergent individuals may exhibit differences in structure and function, they still have the capacity to adapt and create new pathways. If you are neurodiverse, I want you to know that this book is for you too. I see you, and I have considered neurodiverse brains in the making of this book.

If you would like to read more, please see page 313.

Critical periods of development

As children we experience developmental plasticity, meaning that the brain shapes itself as a result of environmental inter-actions. During childhood we undergo the most critical stages of development, moments when the nuances of life are ingrained into our brain maps. These areas in the brain hold different information about us, those around us and the world. Informa-tion must come in for those brain maps to develop, such as love, affection and more nuanced knowledge: your culture and religion, how your family behaves and the places you visit. Ba-sically whatever you experience as a child – the things you see and interact with – shapes your brain, and not one single brain is the same. You are unique and your brain has been built by your surroundings.

Children absorb information like sponges without having to pay attention to it. This is because, during these critical periods, the child's brain is rampant with a molecule called brain-derived neurotrophic factor (BDNF). This molecule is important for help-ing create new synapses and promoting the survival of current ones, which is a key component for rewiring our brains. We will learn how we can access it as adults and what it means for our ageing brain. A key feature that is lost in adult plasticity is that a child's brain absorbs everything and we don't have to actively tell the brain what is important or not. This is why children can learn languages so easily, whereas in adulthood you need to make a more conscious effort. If you put on a French-language recording every day in the background of your child's life during their critical period for language, they would most likely learn to speak French, even if it's not fluently or the whole language. But if you tried this in adulthood, you wouldn't be able to pick it up unless you made a conscious effort to pay attention to the words.[3] So your childhood experiences are important and heavily shape your brain because you were in a constant state of suggestibility. Consequently, these experiences helped create a set of rules about how you should operate in life.

It's important to understand that there are some things that can be changed, some things that require a lot of effort to change and some things that can't. For example, if you haven't developed sight during your critical period, then you won't be able to develop it later because the brain is efficient at using all available space. So if the brain map for sight is available and hasn't been claimed by vision, then it's 'available for rent' and something else will take over. It's why blind people have other heightened senses and have more highly sensitive tactile discrimination for Braille reading than someone who has vision, because they are required to use the other senses more and those brain maps become more developed. You could develop tactile discrimination to learn to read Braille, but you'd need to put the effort in. Things like vision and hearing are extreme examples of what cannot be changed past the critical period, but most of the brain is plastic. Our habits, behaviours, knowledge of the world, intelligence, analytical skills, creativity and other things like learning to dance or playing the guitar – those are all things that we can change.

I asked my Instagram followers what habits and behaviours were holding them back.

HABITS	BEHAVIOURS
Sleep hygiene	Focusing on the negative
Social media usage	Overthinking
Inactive lifestyle and being lazy	Negative self-talk
Procrastination	Insecurity
Food habits	People-pleasing
Grogginess	Lacking purpose
Substance use when stressed	Not speaking up
Perfectionism	Thinking I'm not enough
Not practising skills	Trying really hard to be loved
Self-interrupting when working	Convincing myself people don't care about me

How many of these habits or behaviours are something you would like to change? Luckily none of these traits are aspects of yourself that are hardwired. All of them are modalities that are plastic and can be changed. You don't need to keep talking to yourself negatively or be lazy and overthink your perfectionism. My promise to you is that by the end of this book you'll know how to deconstruct many of these aspects of your life, and you'll be able to apply the knowledge to any scenario so that you can achieve long-lasting change.

We aren't born with low self-esteem
and ruminating thoughts.

As we grow, we learn to attach meaning to things. When you're five or six you may understand that your mother's gold wedding ring is shiny, pretty and somewhat valuable, but you only come to understand what it symbolizes when you are thirteen and the ring is firmly and permanently removed from her left hand. We learn these nuanced and basic behaviours on an unspoken level, often through body language. They can be so subtle it's hard to describe. Have you ever met someone who has the same mannerisms as someone else you know? They remind you of them but you can't quite work out why. It could be in the way they use their hands to speak or a small curl in their lips when they talk. You're not entirely sure what but something reminds you of that other person.

In the Venezuelan tribe of Ye'kuana, mothers constantly carry their newborns for several months after birth so that the babies can learn from them.[4] When the mother reaches for a pot on the stove, she twists her body so that the baby is further away from the heat, and the child learns without having to be told that the stove is hot, so they never go near it in their older years because this understanding has already been encoded into their brain. And so it is with behaviours. Think of how your parents react to particular situations; perhaps your mother is insecure and hates her body, so you gain observational knowledge about how one should feel about appearance and you go on to hold the same beliefs. I remember some of my peers always losing their cool in mildly stressful situations in which they had to deal with conflict . . . because of this, I grew up crying every time I was angry, but it wasn't until I learned about the principles of neuroplasticity that I discovered I could change this.

Observational knowledge is wonderful because we learn that things can burn us just by seeing how our parents interact with the objects around them. But observational knowledge is also how we pick up unhealthy behavioural patterns during childhood. We see how our peers interact with the world and we mimic them, shaping the way we are. We are told by our parents what's right and wrong and we just go with it, because we don't know any different.

What a weird sentiment it is to think that as adults
we have the ability to mould and shape a child to
become who they will be in their adulthood, and most
of the time we don't even realize we're doing it.

Just like hearing an insecure mother complain about her appearance, we also absorb minor remarks or statements that can stick with us for the rest of our lives. How often have you come across people who don't pursue their goals because a teacher told them they would never be good at it? Or parents that attach labels to children like 'they're the sporty one', 'she's not good at maths' or 'she's the creative one; she could never run a business'. Or a parent makes a remark about your behaviour or appearance without malicious intent but it still shapes your belief: women who become housewives because their parents tell them they weren't good at studying and that's what they should do; men who were told they couldn't show their emotions; men who are discouraged from creative work because they should be doing manual work. This subconscious programming feeds in to their own downstream cascade of issues, driven by deep-rooted beliefs. We aren't born with low self-esteem and ruminating thoughts. Have you ever met a child who doesn't say things like 'Mummy, look at me!'?

Ruminating thoughts are a pattern of repetitive, intrusive and prolonged dwelling on the same distressing or negative thoughts, memories or problems. These tend to be focused on past events, worries about the future or self-criticism, and they can be harmful to one's mental and emotional well-being. This is because they intensify stress, hinder our problem-solving skills, disrupt sleep and affect our physical health. They can also lead to social withdrawal, which can further exacerbate negative emotions.

Mental heuristics

While it is harder to make plastic changes as adults, it is not impossible; it just requires a bit more attention and intention. By the time we reach adulthood, we operate on mental heuristics, or mental shortcuts. For example, you most likely don't pay attention to the way you brush your teeth, the way you make your coffee or even which hand you use to open a door or switch the light on or off. But if someone asked you to start doing it differently, chances are that by the next day you would revert to automatic behaviour. Say you brush your teeth with your right hand; it's going to be very hard for you to start using your left hand, and you will most likely forget, because your brain is designed to take shortcuts and use the path that it knows best, to save energy for more cognitively demanding tasks. Your brain is not going to remind you that you need to spend energy trying to use your left hand; that is why it is so hard to break a habit. Most of our behaviours and the ways that we operate in the world are automatic and governed by our subconscious mind, and it takes an immense amount of energy to go against that grain and create new pathways. It's the reason you've had to ask your flatmate multiple times to stop leaving the kitchen door open when they cook; they're not a bad person, and they probably aren't trying to annoy you either – they're just in their own head, operating through mental heuristics that govern the way they operate in the world, completely unaware that they were supposed to close the door for you and forgetting that you'd even asked them to.

Gene McGehee was an elderly man who had dementia. A video of him went viral on TikTok and it moved the entire social media world to tears. The video explained that Gene's memory reset every fifteen minutes, but every day he came out of his home with his chair at a particular time and never really understands why he did this. The video then showed a school bus of children getting off at the bus stop across from his house and running over to him to say hello and shower him with love and appreciation. This put a huge smile on his face every day. These children had been coming to him for three years to have conversations with him, and yet he thought he was meeting them for the first time. And every day he repeated the same pattern, never really knowing why he felt compelled to go outside but deep down something was telling him there's joy out there. His subconscious brain had a pattern of ingrained behaviours and actions that drove him to step outside every day, showing us that we don't always have to be consciously aware of what we're doing for us to perform specific actions in our day-to-day life.

As adults we need to acknowledge what is important to us if we want to learn something new. We need to pay attention to what exactly it is we want to change and what we want to learn. Stay with me. In our brains, we have a system called the reticular activating system (RAS), which filters out all the information that is coming through its sensors (seeing, hearing, touch, taste, smell). For example, if you were in a coffee shop with your friend, you'd most likely be paying attention to the words that they were saying and you would also be looking at them. But without your conscious attention, there are noises in the background and images in your peripheral vision that are still coming through your ears and eyes and communicating with your brain. Your brain is not paying attention to those things because it is constantly filtering out these senses and not bringing them to your attention. At any given moment, there are billions of pieces of information coming into your eyes, and your brain is only allowing the bits of information that are relevant at that moment to be in your conscious mind. You're not aware of this, but your brain is working hard; the busier the surroundings, the more filtering there is going on.

If you wanted to, you could shift your focus to tune in to the sound of the people talking at the next table; attention is being shifted to specialized structures in the auditory cortex, the part of the brain that processes sounds, to tune in to the frequency of sound that is coming from a different place to your friend's voice right in front of you. You could distance that frequency even more to pay attention to the dogs barking outside the coffee shop. If you had to expand your vision to a panoramic view, you could look at your friend but also acknowledge that there are images in the background, perhaps another table or a plant. There are also images in your periphery, but your brain is filtering out all that sound and all those images so that you can pay attention to your friend.

Adult plasticity works in a similar manner. We've touched on why you can't put on a French recording in the background and expect to learn the language like a child could; you'd have to really pay attention to the words, and then through repetition you could learn it. Similarly, if you want to change a habit or unlearn a particular behaviour, you must pay attention to it regularly and then repeat it until it becomes as automatic as brushing your teeth. On average it can take anywhere from 18 to 254 days to form a new habit,[5] depending on the individual. And on average it takes around 66 days for a behaviour to become automatic. The timeline of this may sound demoralizing, but what I am trying to highlight is that these things take time, so if you have tried to make a change but given up by the end of the week, it's because your brain is reverting to what is automatic and mental heuristics.

A friend of mine once told me that I used to complain a lot, and that this was why he and his girlfriend never wanted to hang out with me. Yikes! That was a bit of a shock. But, you know, it was clearly what I needed to hear. That behaviour was so deeply ingrained that I barely even noticed I was doing it; it was just an automatic way of operating through life – constantly complaining, constantly huffing and puffing as I walked around the house. But as soon as he said that, I started paying attention to every single time I complained, which meant that I could put in the work to change the behaviour. Sometimes people want a grander solution, but acknowledgement is step number one. You can't grow a plant without planting a seed, and planting a seed is acknowledging that you have a problem that needs changing.

This book is going to help you break the cycle, alter your thoughts and create lasting change through three phases. In Phase 1, *Ditch the Negative*, we'll explore how to let go of anything that's holding you back. You'll learn about stress, good and bad, and how to regulate your central nervous system, and this is where we will lay out the fundamentals of how our brains function so that you can put in the groundwork needed to make a change. You'll learn why our brains like to focus on the negative and how to change that. Phase 2, *Shift Your Narrative*, is where you'll put in the real work to change your brain's hardware, and by following the seven steps to rewire your subconscious, you'll learn how to change your own narrative for the better. In Phase 3, *Boost the Positive*, you will learn about how to support your hardware for long-lasting change. This final phase will help you understand how to reach peak levels of well-being and maintain them on a day-to-day basis.

I'll be here with you on this journey, and by using evidence-based science and real-life examples, I will teach you how the brain is capable of change and what that can look like for you in real terms. This book is an interactive toolkit that is structured in a way so that you can get the information you want from the first section of each chapter, with a scientific follow-on for anyone who wants to delve deeper.

'You are the most wonderful project you'll ever work on' – Unknown

Phase I

Ditch the Negative

Dwelling is learning. Dwelling teaches us. Let us dwell. But only for a short while.

- Break the cycle
- Negativity bias
- The power of your thoughts
- Creeping normality
- Confirmation bias: you'll see it when you believe it
- Endings, loss and grief
- Neuro toolkit 1: how to ditch the negative

Break the Cycle

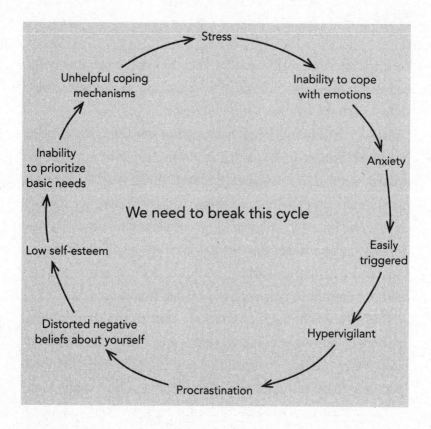

Stress

Unhelpful coping
mechanisms

Inability to cope
with emotions

Inability
to prioritize
basic needs

Anxiety

We need to break this cycle

Low self-esteem

Easily
triggered

Distorted negative
beliefs about yourself

Hypervigilant

Procrastination

When we're in a rut it can be hard to see a way out. Even when we have all the answers, it's hard to believe that things can change. This is down to several contributing factors, from the state of your neurobiology at any given time, which drives your mood, unsolicited emotions and your stress levels and thoughts of worry.

Stress

Let's tackle stress first because, let's be real, often this is the culprit for keeping us stuck. Stress can be so confusing, and rightly so, but I will break it down for you.

Stress is a term that often has negative connotations, evoking images of tension, anxiety and discomfort. However, not all stress is detrimental to our well-being. Stress is necessary in many circumstances, and by understanding its diverse roles we start to understand how we can get to grips with it and manage it better. This is important because sometimes we attribute stress to situations that are distressing but not exactly stressful, and this has the potential to perpetuate a self-fulfilling prophecy.

There's a whole body of literature that shows that our perception of stress, and changing the way we think about our stress response, can improve our physiological and emotional reactions to stressful events.[6] Participants in one study[7] were told that their physiological response to stress was functional and adaptive. During the testing, their heart rate and blood pressure were lower during a stressful task relative to the control group, showing that the perception of stress as good can alter the body's response to it. By adopting this mindset you can foster a more logical and realistic outlook on stress, which will allow you to assess your stress levels accordingly.

I break stress down into three categories: altertness, acute and maladaptive.

Alert stress

Contrary to popular belief, stress is a requirement for heightened performance and focus. On many occasions you need some level of alertness to perform at work, during exercise and in daily activities, such as walking your dog and doing your chores. What this means is that there are hormones such as adrenaline and cortisol, as well as neurochemicals such as norepinephrine,* circulating through our body to prepare us for these events. These chemicals are responsible for enhancing alertness and even improving reaction times. A moderate level of stress is often associated with optimal performance, as it motivates us to meet deadlines, tackle problems head-on and achieve our goals.

I believe that we often use the word 'stress' to describe how we're feeling when, in fact, we may be in discomfort. I think it's an important distinction to make because the blanket statement 'I am stressed' diminishes our locus of control over how we're feeling. When we properly assess our situation and

---

* You may see norepinephrine and noradrenaline being used interchangeably throughout this book. They are the same thing, just spelled differently in the US and UK.

appreciate that we're in a place that requires us to be ready for whatever activity we're engaging in, then we can start to manage our stress better.

Acute stress

Acute stress is adaptive; it is designed to be a short-term burst of energy that aids in problem solving and coping. It's adaptive because after experiencing a stressful event, the body adjusts to that scenario to get you out of it and so it promotes personal growth and resilience. For example, the first time you had to prepare for an exam, you experienced a heightened level of stress. This stress response would have triggered you to increase your focus, alertness and motivation to study effectively. The temporary surge in stress would have aided you in managing your time more efficiently, prioritizing tasks and implementing effective study strategies. This would have taught you coping mechanisms for dealing with stress and time-management skills, and increased your resilience in the face of pressure. This would have prepared you for future academic challenges. It's why I stopped worrying about my MSc coursework deadlines, whereas in my BSc I experienced a near meltdown every time I had something to hand in. By the time I got to my MSc, I stopped viewing these deadlines as impending doom and instead as an opportunity to get the work done in a more orderly fashion.

Stress plays an adaptive and positive role in facilitating personal growth and success in challenging situations. When managed appropriately, acute stress contributes to heightened productivity and effective decision-making.

Life is full of challenges, and adapting to these challenges often involves a certain degree of stress. Overcoming obstacles can lead to increased mental and emotional strength, fostering a sense of accomplishment and self-efficacy. Adaptive stress encourages individuals to develop coping mechanisms, problem-solving skills and a greater capacity to navigate future difficulties. It acts as a transformative force, shaping individuals into more resilient and capable beings.

Maladaptive stress

While arousal and adaptive stress can be beneficial, maladaptive stress represents the dark side of stress when it becomes over-whelming and harmful. This type of stress, occurs when the demands placed on us exceed our ability to cope effectively. Mal-adaptive stress is when we're mentally, physically and emotionally exhausted, unable to meet the demands. It starts to compromise our well-being and can contribute to a range of physical and mental health issues, including anxiety disorders, depression and cardiovascular problems. This usually happens when acute stress doesn't go away and it becomes chronic, or when there are too many demands and not enough stress reduction.

Dealing with chronic stress that is maladaptive can push us to our limits. When this happens, our brain and body turn on a 'safe mode' of living, or 'low-power mode'. Just like when the battery on your phone is running low, low-power mode saves energy for the most basic functions and lengthens battery life. When you're in this mode of living, your brain reverts to mental heuristics, those shortcuts you're so badly trying to avoid, meaning you repeat old habits you're trying to shake. Decision-making is impaired and your brain will take the route more travelled. Why would you run down a dirt path if you can just drive on the motorway when you're exhausted? When we're living like this, the brain cannot prioritize making habitual and behavioural changes. The hardware that is your brain is working hard to operate and it cannot make any software updates without crashing.

Key areas in the brain that are affected by chronic stress include the frontal cortex, the amygdala and the hippocampus.

The frontal cortex:

- The frontal cortex is involved in higher order cognitive functions such as decision-making, problem solving and impulse control.
- Chronic stress can lead to structural changes in the frontal cortex that may impair the brain's ability to regulate the amygdala, which is involved in emotional processing, and therefore disrupts how we respond to stressors by making us more hypervigilant.

The amygdala:

- The amygdala is a key region involved in the processing of emotional responses, including fear and stress.
- Chronic stress can lead to an increase in the size and activity of the amygdala. This heightened activity may

contribute to an exaggerated emotional response to future stressors.

- This overactive stress response can impact emotional well-being.

The hippocampus:

- The hippocampus is crucial for memory formation and the regulation of the stress response. It plays a role in shutting down the stress response after a stressor has passed.
- Chronic stress causes structural changes to the hippocampus, which can contribute to difficulties in learning and memory.

This is important knowledge because it helps us tackle acute stress and chronic stress accordingly. It's valuable to understand why you're feeling overwhelmed and hypervigilant, with the inability to cope. But the brain is plastic and this can change. When you're in low-power mode you may find yourself being more easily triggered and irritable due to the brain changes you've experienced that have been caused by stress. Or perhaps you're lethargic and tired all the time yet you've attached that feeling to laziness so you berate yourself for not getting any work done while pushing yourself further into this rut that you can't get out of. So if you've been struggling to concentrate at work or finding yourself getting ill all the time and apathetic towards life, you've probably reached your limits. You're not lazy and unmotivated. You're unable to cope with life's demands. Your brain is not going to prioritize getting your homework done or preparing for that meeting. Instead it's going to prioritize trying to replenish energy resources in any way that it can. If you think of your brain and body as a system that requires energy for every cellular process – from thinking all the way through to running – then if your energy resources are depleted due to constant stress, something must give.[8] Your homework is the first to go. Choosing delivery over cooking is probably second.

I feel as though I lived most of my twenties in a chronic state of emotional dysregulation and maladaptive stress. I was always getting sick, I was drawn to junk food quite often because of my inability to prioritize cooking for myself and, quite frankly, I was a mess. They don't call them the Roaring Twenties for nothing. And boy was I roaring.

Of course, being in this state isn't only limited to your twenties; we can encounter this level of stress at any point in life. I sure went through it a few times again. But, for all my sins, the worst bout of it, which manifested itself in an inability to cope with life, came during my biggest era of self-discovery. Double whammy. Exercise felt hard and I was easily triggered by innocuous events and often irritable. Fun times. But, you know, you do the best with the information you have at the time, and I certainly didn't know any of this then. Hence why I am here, passing on the baton of information to those who are lucky enough to receive this before finding weird and creative ways to get themselves in a mess like I did. Had I known then what I know now, I would have learned to manage the hustle and bustle of London life a bit better. There were financial worries and some toxic friends, coupled with unresolved childhood trauma and garnished with regular partying, which wasn't limited to weekends by the way – the cherry on top of all this mess. Wednesday hangovers after Tuesday nights at G-A-Y, drinking £1.80 beers, were a regular occurrence. Sorry to any of my previous bosses who are reading this and believed my lies about 'being tired'. But nobody has a degree in hindsight. I do, however, have one in neuroscience, so that's a win. I now understand that my unhealthy coping mechanisms led me to a place that was increasingly dark because I didn't have the tools to manage my own emotions. And I certainly didn't know enough about stress to even understand what I was feeling or going through. On top of that, I made lousy choices because I didn't have the self-esteem to advocate for myself, so I ended up in desperate romantic and platonic relationships, hoping for someone to cure me.

I had a lot of work to do with having to rewire self-beliefs, but for that to happen I needed to go back to basics and learn to cope with my stress better. I was living in a state of constant stress for a variety of reasons but also because my basics were not in place. When we think about the brain as the hardware, the basics such as sleep, not drinking alcohol, learning to meditate (among many of the tools that this book will give you) and setting clear boundaries about what you want or don't want will

REMOVED THE EXCESS & FOCUSSED ON ME

REMOVED UNNECESSARY LOAD FROM MY LIFE (EXTRA WORK, SOCIAL MEDIA, LARGE GATHERINGS, ETC.)

PRIORITIZED WHAT WAS IMPORTANT; LET GO OF WHAT WASN'T

COMMUNICATED MY NEEDS WITH FRIENDS AND FAMILY

LEARNED TO SAY NO WITHOUT GUILT – BOUNDARIES

CREATED CLEAR WORK/HOME TIME SCHEDULES

AVOIDED ALCOHOL AND PARTYING

MEDITATION & SELF-HYPNOSIS

HOBBIES – BALLET & ART

REGULAR EXERCISE

LIMITED CAFFEINE

TIME IN NATURE

GOOD SLEEP

YOGA

ensure that the machinery is working correctly so that you can make upgrades to your software.

My first realization was that perhaps it wasn't the world that was conspiring against me. I was conspiring against me.

HOW I BROKE THE CYCLE

The stress mechanism

The physiological reaction of stress is orchestrated by the amygdala, a key player in the brain's limbic system. Acting as a command centre, the amygdala swiftly assesses environmental stimuli, particularly those perceived as threats, and when it detects danger it sends signals to activate the body's stress response through the hypothalamus–pituitary–adrenal (HPA) axis to activate the **sympathetic nervous system**.

The amygdala's role is pivotal in triggering the release of stress hormones, primarily adrenaline and cortisol, from the adrenal cortex. This process is crucial for preparing the body to face challenges by heightening awareness, increasing energy and sharpening focus – commonly known as the fight-or-flight response. The amygdala's ability to rapidly process information allows for quick decision-making in potentially hazardous situations.

In contrast, the **parasympathetic nervous system** acts as a counterbalance, promoting a rest and digest state. The parasympathetic system plays a crucial role in restoring the body to a state of equilibrium after a stressor has passed. Key to this calming influence is the vagus nerve. The vagus nerve interfaces with various organs, including the heart and digestive system, and releases acetylcholine, a neurotransmitter that counteracts the effects of stress hormones. When activated, the parasympathetic system slows the heart rate, enhances digestion and promotes relaxation.

Balancing the activity of the sympathetic and parasympathetic systems is essential for overall well-being.

Stress hormones in the brain and body are molecules that are uncoupled to the cause of stress, meaning that on a fundamental level the brain perceives stress regardless of the source or the event, and it cannot differentiate between the stress of a looming deadline at work and the stress of being chased by a lion. The magnitude of the stress response can vary, which can have a more significant impact on the brain and body, but otherwise it's a generic molecule that communicates what to do in a specific scenario with your brain and body. Run, focus, fight or finish that report.

We all feel stress, some days more than others, but usually we recover from that stress shortly afterwards or within a day. As humans we are supposed to have a stress response; it's healthy and necessary. You will always encounter stress; it's naive to try to live in a world where we avoid it. The key to dealing with stress is to reframe our mindset towards it, but more importantly to recover from it quickly so that we don't linger in a state of stress and risk it becoming chronic and maladaptive. This can be hard as we often experience continuous stress from things that just won't go away, like divorce, money concerns or health issues, and that's when stress can start having detrimental effects. Chronic stress can cause brain changes which can contribute to depression, anxiety and other problems. This can leave us with an inability to regulate our emotions and lower our tolerance towards common, everyday stressors. This is because the amygdala is overactive and can begin perceiving threatening stimuli in the environment that would not normally cause a stress response. You know that feeling when you cry because your jeans got caught on the door handle? Mundane but also so valid.

I was there for the most part of my life, as a product of a turbulent household that didn't teach me how to deal with my emotions, or how to live freely without the outbursts and reactivity I had observed of my family. I bet my little brother is laughing his socks off reading this, knowing what I used to be like. A prime example is the time I had a massive meltdown with him outside a car rental place in the Italian Alps. I hadn't read the Ts&Cs (does anyone?) and didn't realize I needed a credit card to rent a car. Which I didn't own. And because I had such a limited ability to regulate my emotions – a testament to the hypervigilant state I was in – that relatively minor inconvenience resulted in a relatively outsized breakdown. I was trying to search for something in my bag, which for the life of me I can't remember now, throwing all my stuff out of my bag. It definitely didn't require that level of drama, that much I know. Yet a mere ten minutes later, after coming to the reasonable solution of putting down a bigger deposit on the car, I looked at my brother while driving, recognizing his bewilderment at my over-the-top performance, and just burst out laughing. 'I'm a star, baby.' It's funny to reflect on but now I realize how turbulent my mental state was. Yo-yoing from despair and frustration to uncontrollable laughter within the space of an hour. And over an issue that turned out could be easily fixed. It's safe to say I was a stressy messy. But by learning the neuroscience behind stress and our physiological response to it, I finally had the tools to manage my life and get myself out of this mode of living. To make any behavioural changes, we must manage our stress response. Mine was shot; any little trigger sent me off into a frenzy . . . evidently.

Living in a constant state of alertness, reactiveness and readiness to attack means that your brain is not going to prioritize making new connections to rewire your narrative. If your brain and body are flooded with stress hormones, it means that the hardware component needs to deal with the stress before you can look at the software. Learning to respond to your stress accordingly means you can focus on getting the hardware working properly again, so that you can then tackle the software of your thoughts, beliefs and emotions. When you're in low-power mode, your brain will prioritize basic functions instead of trying to change your habits and behaviours. It's hard to make any plastic changes when you're maladapted to stress.

OK, so the following is what is supposed to happen when we experience an acutely stressful event.

Let's say that you experience a stressful event. There should be a parasympathetic rebound from your body, signalling to your brain and body that the threat has gone, allowing you to return to baseline. Sometimes this doesn't happen because we have problems that don't go away, and sometimes we continue to linger on a problem by thinking about it. The highly developed human brain has the ability to continue thinking about stressful events, which keeps our cortisol levels raised. Our brain and body continue to perceive being under threat, and this is maladaptive because it is not providing any function. So what are you supposed to do?

Mild acute stress reaction

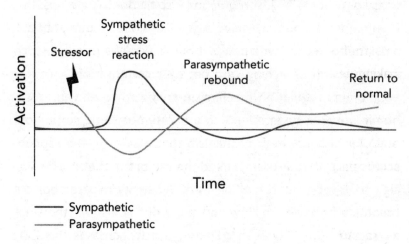

Adapted from: Payne, Levine & Crane-Godreau, 2015. *Front. Psychol.*[9]

Tool 1

The brain is very primitive and it's not designed to listen in a time of threat and danger; it's designed to run, fight or freeze. This means that if you want to shift back into a relaxed state, your body needs to communicate with your brain that the threat is no longer being perceived. The quickest way to communicate this to the brain is through a breathing technique which neuroscientist Dr Andrew Huberman has coined as the 'physiological sigh'.[10]

Have you ever felt stressed and overwhelmed and you started walking around sighing or huffing and puffing? This was something I realized I did when I first got my dog Kobe. As a Border collie, he's ultra-sensitive to emotions and energies, and he used to hate it when I would sigh. I didn't realize how much I did it when I was stressed until I got Kobe. We now know each other very well and he's learned to understand that I am not angry or upset – I'm just regulating – so he doesn't get upset by it any more. If you don't have a dog like Kobe, you might have a partner or a child who points it out to you. We usually associate the sound of sighing with negative connotations, but it is your brain's regulation mechanism trying to reset your autonomic nervous system back to a parasympathetic state. Your brain has a hardwired mechanism that resides in the subconscious part of your brain, called the midbrain, that makes you sigh to reset your breathing rate. When we're stressed, our breathing becomes shallow and some of the alveoli (the small air sacs) in our lungs can collapse, which increases the CO_2 levels in our blood. The abundance of CO_2 causes us to feel stressed and agitated, which is why taking deep breaths can feel calming. Sighing has an important function in ventilating the lungs by allowing for maximal expansion, which prevents any further collapse of the alveoli. Thus sighing is a natural part of our behavioural response to stress. Now that I've brought it up, pay attention to how much you sigh when you're stressed and anxious . . . or perhaps someone has pointed it out to you already. With an understanding of the benefits of this sigh, you could explain to them why it's such an important physiological response and a vital part of our breathing mechanism.

You can also fast-track this mechanism by deliberately forcing it. 'The physiological sigh', sometimes referred to as 'cyclic sighing', is a breathing technique that can fast-forward this automatic mechanism by inducing a calming response manually, and it has been proven to be the most effective way to immediately bring you back down into a parasympathetic state.

The double inhale is important because it helps reinflate the collapsed alveoli by forcing them open again. This allows for more of the lungs' surface area to increase the intake of oxygen and efficiently remove more CO_2 from the system, which signals to your brain that you're no longer experiencing anything threatening. This allows your heart rate to come down, which in turn brings you back to a relaxed state.

Huberman stated that the physiological sigh could serve as an anxiety-reducing tool by giving the individual autonomy and control. One of the biggest hallmarks of anxiety is the perceived lack of control during a situation, and the physiological sigh empowers the individual to take control of it by fast-tracking the stress response back to a state of calm.

This is why the physiological sigh is a great tool in helping us manage our stress response. When we focus on lengthening the out breath, we are signalling to the brain that the body is no longer in danger and the stress response can be 'switched off'.

HOW TO DO THE PHYSIOLOGICAL SIGH
TAKE TWO SHORT INHALES
(IDEALLY THROUGH YOUR NOSE, BUT IF YOU
CAN'T, THEN THROUGH YOUR MOUTH IS FINE)

PAUSE FOR ONE SECOND

EXHALE LONG AND SLOW THROUGH
YOUR MOUTH

Tool 2: Hobbies after a long day

In the introduction we learned that our thoughts can evoke physiological responses, and we saw how our brains can still perceive stress by thinking about a stressful event. If we're ruminating over something that happened in our day, the brain can still perceive the event to be taking place even if we've removed ourselves from it, and therefore if we come home after a stressful day and continue to think about work, our stress hormones don't subside.[11] By engaging in hobbies, we immerse ourselves in something else so that our mind doesn't think about these stresses and it has a chance to regulate in that time of distraction. Hobbies can be activities that you regularly engage in, like singing and dancing, or perhaps you're a musician. They can also be pastimes for pleasure and relaxation, and they're usually done during your leisure time. Ideally you'd embark on a hobby that takes your mind off things, something in which you can become completely absorbed, allowing your brain to recover from the stressful day you've had and absorb all the problems in your life.

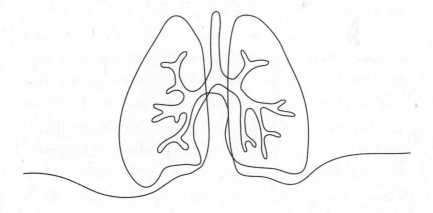

Additionally, engaging in hobbies has a profound impact on our mental well-being. Hobbies make us feel good after we've engaged in them because they stimulate various regions of the brain that challenge us, and this causes a rise in BDNF. BDNF is released through activities like exercise, learning and creative pursuits associated with hobbies.[12] The increase in BDNF boosts our mood, reduces stress, enhances overall mental health and contributes to a sense of happiness and fulfilment.

Engaging in regular hobbies will allow you to come home from work and leave your stress at the door so that you can enjoy your evening with your loved ones without ruminating on the day's events. We're never going to be able to avoid stress, and we shouldn't try to either; we should embrace it because acute stress is what keeps us alive and functioning, but we need to understand how to live a life where we aren't constantly tipping the scale in a negative direction.

The issue is when stress becomes chronic and doesn't go away. You're stressed at work due to an argument you had with a colleague and then you come home to a failing marriage where you're figuring out how to go through with a divorce; you then drink wine in the evening to try to dampen the thoughts, and now your body is constantly flooded with stress hormones without a chance to recover. This means that you remain in a sympathetic state that is maladaptive, and your ability to come back down into a parasympathetic state is compromised. Overall the system is disrupted so that the scale is tilted negatively towards a sympathetic state that operates in excess, and the parasympathetic state is not responding appropriately. If this state of chronic stress isn't dealt with, it could persist indefinitely, and the brain and body could start to respond inappropriately to environmental challenges due to this excess activation.

This is what a state of chronic stress looks like.

You can see from the graph below that the parasympathetic system is incapable of recovering back to normal. Even a moderate stressor is causing an overreaction in the stress response. We call this allostatic load, and it describes the cumulative wear and tear on the body as a result of ongoing or repeated stress. It reflects the physiological cost of dealing with constant cumulative stress.

When you're experiencing chronic stress, it's hard to concentrate at work or perhaps you're even struggling to take care of yourself. You find it hard to get the energy to even shower. This

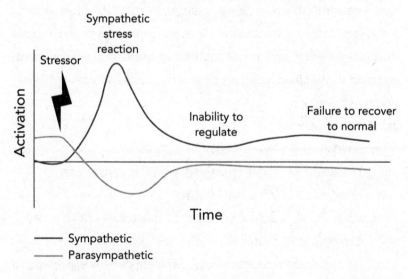

Chronic stress response

Adapted from: Payne, Levine & Crane-Godreau, 2015. *Front. Psychol.*

is because your brain and body are in low-power mode. If you're hypervigilant and easily triggered, your emotional state may be heightened due to your levels of stress, leaving you feeling overwhelmed and reactive towards things that usually wouldn't trigger you (like not having a credit card when renting a car). When you're in this state of chronic stress, your brain is not going to prioritize trying to make changes to your way of thinking. This is why you have felt stuck in this cycle.

Tools for chronic stress

Managing your stress is also crucial for emotional regulation because the two are closely connected. Having a nervous system that responds properly to stressors allows you to become more attuned to your emotions without being overwhelmed by them. This heightened awareness helps you identify and understand your emotions, making it easier to regulate them effectively. When the nervous system is activated in response to stress or negative emotions, it can lead to an escalation of emotions; by using these tools, you can intervene early and prevent emotional escalation.

Take-home messages:

1. Stress is not always bad.
2. The word 'stress' is overused, which means it can undermine our mental health.
3. Learning to regulate your stress shortly after the event can help you manage it.
4. Our mindset towards stress can change how we adapt to stress.

TOOLS FOR CHRONIC STRESS:
GO BACK TO BASICS

SLEEP
THE BEST RESET BUTTON FOR STRESS. ALLOWS YOUR BODY TO RECOVER.

MEDITATION/SELF-HYPNOSIS
HELPS SHRINK THE AMYGDALA. INCREASES BRAIN PLASTICITY. IMPROVES RESILIENCE.

REGULAR EXERCISE
RELEASES BDNF, WHICH IS POTENTIALLY ANTI-DEPRESSIVE. LOWERS INFLAMMATION. RELEASES ENDORPHINS.

REMOVE THE EXCESS
CREATE BOUNDARIES AND REMOVE THE THINGS THAT TAKE YOUR ENERGY AWAY FROM YOU. SAY NO MORE.

HOBBIES
IMMERSE YOURSELF IN AN ACTIVITY THAT MAKES YOU FEEL GOOD.

Brains and emotions

You know that feeling when you're super emotional and you can't think straight? You act on impulse and you're overburdened with emotions that feel as though you'll never get them under control? That's because our emotional responses are deeply connected to various networks in our brains that govern different cognitive processes and which can therefore influence our ability to perceive and process information. Emotions are necessary for communicating what is going on in and around us, in our internal states and the external world. And while we don't have complete control over the emotions that arise, we actually have a lot more control over them through emotional regulation than we believe. I feel as though we've been taught as a society that emotions are bad and we shouldn't feel them, but you're not 'too emotional' for having them. In fact, all these emotions are what make us human. So by learning what is going on in your brain when you're experiencing intense emotions, as well as learning how to manage them, you can learn to deal with them instead of trying to push them away and pretend they don't exist.

Some people hold within themselves the narrative that they're a sensitive soul or that they're an emotional wreck who can't control their emotions. I have been told before that I shouldn't use my emotions when making decisions, and while to some degree this can be true – for example, you don't want to make impulsive decisions when you're angry – emotions are a vital part of decision-making and there are very few decisions we make in life that aren't accompanied by some form of emotion.

The problem is that when we feel overwhelmed by our emotions, or when the emotions are intense, they tend to consume our mental resources that we would usually use for higher-order thinking (problem solving, creative and critical thinking). When we're overwhelmed by emotion, our brain's capacity for complex reasoning and decision-making may be impaired, depending on the magnitude of the issue. Strong emotions can draw our attention away from other things too. Say you get a dreadful text from your partner telling you that they want to have a chat when you get home from work. You know the relationship hasn't been great for a while, so now you become fixated on trying to figure out what is wrong. You may experience intense feelings of fear and insecurity, which can then shape the rest of your work day by making you interpret events at work in a way that aligns with your emotional state. Maybe your boss calls you into their office to praise something you did, but your bias from the insecurity you're carrying distorts the way you approach their office. You may even feel as though your boss is being sarcastic when they're being nice or you're waiting for a catch that never comes.

Emotions can have a significant impact on our ability to think clearly. They can also influence how we retrieve memories, leading to biases in how we recall information, which can skew our perception of events. I know I have certainly felt this when

I've been in an argument with my partner and all of a sudden things that don't usually bother me rise to the surface and are skewered by my emotional state. So that one text he sent on 23 July 2018 has spontaneously become relevant to today's fiasco. I'm exaggerating but you know what I mean. Hopefully I am getting a chuckle or a nod at least.

What's going on in your brain

The human brain is a highly interconnected and complex organ with an amalgamation of different regions working together to perform various functions. However, these regions have their own functions and often operate with different characteristics.

The **limbic system** is responsible for motivation and emotional processing and regulation, and for triggering the fight-or-flight response. It's often referred to as the 'emotional brain' and it influences our responses to different experiences and sensations. Information processing in the limbic system is mostly subconscious and it is impulsive and fast; it's designed to prioritize immediate survival over rational thinking. When intense emotions are triggered, they can hijack our more logical and precise counterpart, the level-headed cool-guy area in our brain called the **prefrontal cortex**.

The prefrontal cortex is responsible for executive functions such as focus, attention and self-control, but more importantly, processing in this area is slower and more calculated. It's because of the prefrontal cortex that we are capable of rationalizing certain scenarios, understanding events that unfold around us and adjusting our response for the future when we make mistakes. This part of the brain is also called the neocortex, 'neo' meaning new. Our developed prefrontal cortex is what differentiates us from other species because we have the ability to ponder our life, judge and have free will. The ability to examine and think about our own thoughts is called metacognition. As humans we also have the capacity to mentally place ourselves in different times of our lives, also called autonoetic consciousness. I'm telling you all this because I want you to have a grasp on how wonderful the human brain is, and for you to see that you have a lot more control over your thoughts and behaviours than you may have previously believed.

The limbic system and the prefrontal cortex can often feel like they're in conflict. Should I? Shouldn't I? The classic tale of 'my mind is telling me no but my heart is telling me yes'. Confusion galore. But it's by understanding the interplay between these two areas that we can start to unite them so that they can complement one another. These areas of the brain, even though they have different roles, are not mutually exclusive and they can work together to shape our understanding of the world and our behaviours. It doesn't have to be you against you all the time. That being said, in some situations the emotional brain can override and influence our sense of logical thinking, usually when emotions or stress levels are heightened. It's the reason stress takes over, because the brain is concerned with survival and thus logic can be impaired. If you think about it, your brain is designed to get out of a stressful situation, not to make sense of it; this makes it difficult to consider alternative solutions or think creatively. If you were being chased by a lion, you would be solely concerned with getting yourself out of that situation instead of trying to understand why there's a lion in your vicinity.

Your brain is mostly concerned with self-preservation, fun later, and it'll do everything it needs to make sure you stay alive. Only when those needs are being met can executive function and other more cognitively demanding tasks such as problem solving and time management take place.

What about strong emotions?

Have you ever made an impulse decision, screamed at your friend or partner, or done something you're not proud of in the midst of stress and despair? That's because your decision-making abilities have been impaired by the limbic system. Under normal circumstances emotions play an essential role in shaping our understanding of the world and helping us navigate it. Our emotions provide valuable information that guide our responses to various situations. Some days you'll be able to be more logical about something, and other days you may cry because you left your water bottle at home. When we're highly stressed, tired or in a heightened emotional state, our limbic brain takes over and logical thinking and executive functions go out of the window. Have you ever tried to problem-solve when you're highly stressed and anxious? You tend to ruminate and go round in circles, unable to find a solution or think clearly, second-guessing yourself.

In the stress section (page 37) we learned about the amygdala overriding the logical centres in the brain, and that's exactly what is happening here. Because your limbic brain includes structures such as the amygdala and the hippocampus, both involved in processing emotions, your perception of certain situations may be altered. When strong emotions such as fear and anger are triggered, they can influence our decision-making processes, putting us in a state of heightened emotional awareness, whereby our thinking is biased and rational thought is overridden by negative and overwhelming emotions.

You know that hole you go down where all you can focus on is the negative in your life? The hole where you feel like you want to run but you have nowhere to go; you want to get out but you don't even know where to begin? I know that feeling, and there have been times when my mental state was so poor that no matter how much knowledge I had on these topics, I felt as though none of it was going to work.

We spoke about anger having a major effect on our emotional state and overriding our logical brain. When we're in these elevated emotional states, our logical areas are essentially being hijacked by the limbic system to keep us alive. Have you ever been in an argument where you've completely exploded and you can't even recall what was going on around you? It's as though nothing else exists and the argument completely takes over – you're seeing red and you say irrational things. Impulse control is compromised and it's dictated by the rampant amount of noradrenaline that's flooded your brain. Your emotions are so heightened that your rational thinking has gone out of the window. When we're feeling extreme emotions of rage and anger, our limbic system is highly engaged and it's hard to think clearly. This is because activity in the amygdala is heightened and this downregulates the frontal cortex, making it harder for you to think logically.

Tool 1
The physiological sigh (page 41) is a great tool in this scenario to try to reinstate some prefrontal cortex function.

Tool 2
Stepping outside into the light and looking out at a panoramic view can help alleviate the feeling that you have no control over your emotions. By widening your field of vision, you can reduce the tunnel vision that's associated with roaring anger.[13] Going outside and taking a few deep breaths, perhaps including the physiological sigh, while looking out into the distance could help mitigate some of the overwhelming feelings you get when you're angry. It also affords you a few minutes to get distance from the scenario and your environment so that you can get clarity. This means you're giving your logical brain some time to think and process so that you don't respond with your limbic brain and spout things that you'll regret later.

Managing all other emotions
I want to acknowledge that I completely relate to that feeling of not knowing if things will ever get better, but I have also come out the other side.

I want to emphasize that emotions are important and we should not try to avoid them. In fact, having a good balance between emotions and logic is a desirable state for living. We just need to make sure that the scale isn't considerably tilted towards an emotional state so that you're feeling stressed, anxious and constantly overwhelmed all the time. And, more importantly, that you're not biased by negative emotions which then shape your life and your narrative.

Studies[14] show that developing emotional intelligence can help us understand and manage our emotions better. By learning about our emotions and about stress, we learn to be more flexible in how we respond to situations and thus we're able to make more sense of why we may be feeling a particular way. This means that emotions can have less of an impact on us or that we're able to respond more appropriately. Alongside this, learning to regulate our stress response can also lower the magnitude of some of the emotions we're experiencing so that they don't 'blow up out of proportion'. When you're extremely stressed, your emotional awareness is heightened and something that wouldn't usually bother you sparks off a cascade of tears. You know that classic millennial meme about your hair tie snapping being the final straw of your sanity? Yeah, that feeling. When you're just about keeping it together and that's the thing that sends you into a roaring rage of tears and emotions . . . Well, clearly there's a lot more going on there, but that was just the cherry on top.

Learning to identify and accurately name[15] our emotions can provide us with clarity about how we're really feeling and also help us become more emotionally aware of ourselves. Studies[16] show that even just naming our emotions can give us a sense of control over them, putting us in a better position to deal with them. This is because naming our emotions recruits our frontal cortex, the area responsible for logic that is usually quietened during heightened states of emotion. This gives us back control and puts us in a state where we're able to respond better without the emotions taking over.

There are many models and charts that researchers and therapists use to help people identify and express their emotions, which can be difficult to categorize neatly as they often involve a range of physiological, cognitive and subjective experiences. One of these models is the Wheel of Emotions devised by Dr Robert Plutchik.

The eight basic emotions that Dr Plutchik proposes are highlighted in the boxes. Within and between the basic emotions are other emotions that we may feel. You can use this wheel to identify what it is you're feeling, to give you more autonomy during emotional times. Having emotional awareness is a valuable tool because it helps us detach from the narratives we repeat in our heads. When we are able to understand our emotions, we are better equipped to deal with the way they trigger the negative stories we tell ourselves based on these emotions. This also helps us see and understand that we may be talking to ourselves in a particular way because of the emotional state we are in, not necessarily because we are speaking facts.

The Wheel of Emotions

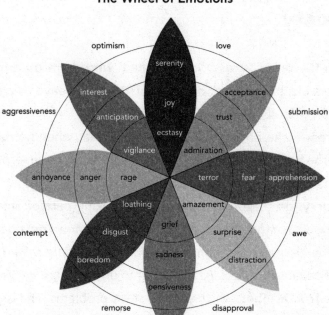

Dr Plutchik 1980

Often we find ourselves ruminating and repeating negative self-talk when we're in a state of chronic stress or when our emotional brain has taken over our logical centres. This solidifies a network of areas in the brain that communicate with one another negatively to shape your entire life.

Remember that emotions are not your enemies, but rather they are the messengers of your inner world. The inner universe that only you'll ever see. They provide insights into your desires, fears and needs.

Anxiety

Often the terms 'stress', 'anxiety' and 'worry' are used interchangeably, but the lack of differentiation can leave us unable to appropriately deal with how we're feeling.

Stress is the physiological response to fear when hormones flood the body, getting us ready to run, fight or freeze in response to the danger.

Worry, on the other hand, is the thinking part of anxiety, when we have thoughts such as, *I don't know how I am going to make ends meet this month,* or, *Did those people think I was weird because I said x, y, z?* Thoughts of worry are generated by the frontal lobes, which want to solve problems. The issue is that these thoughts can become repetitive and ruminative, which can turn into anxiety.

Anxiety is triggered in the limbic system, the emotional part of our brain. Usually the amygdala acts as a threat-detection system, which keeps an eye out for potential danger; if there is danger, then the amygdala signals to the brain that stress hormones need to be released so that one can be prepared to deal with the imminent threat. However, sometimes this response is overactive and signals danger when there isn't anything to be fearful of. When faced with an anxious situation, the brain's attentional control system tends to prioritize threat-related stimuli, diverting cognitive resources away from constructive problem solving. As a result, individuals experiencing anxiety may struggle to redirect their attention towards more rational and solution-oriented thinking.

Worry

- Worry is a cognitive process that involves thinking about potential problems or situations that may or may not occur in the future.
- It usually involves a sense of unease, uncertainty or fear about what might happen. Worrying is often focused on specific events or outcomes.

Anxiety

ACUTE STRESS	ANXIETY
Short term	Can linger
Triggered by something external	May not have an identifiable trigger
An external pressure that's causing you to respond accordingly	Does not disappear when external stressor is removed

- Anxiety is a more general and pervasive emotional state characterized by a sense of apprehension, nervousness or fear. It is often accompanied by physical symptoms such as increased heart rate, muscle tension and restlessness.

- Anxiety can be a response to a real or perceived threat, and it may not always be linked to a specific event. It can manifest as a chronic condition or as episodes of intense anxiety.

So what about stress and anxiety?

Anxiety and stress are often used interchangeably, but while they are different, they can frequently overlap and interact in various ways. I often hear people saying that they're stressed about something when, in fact, they are worried about a potential future outcome. They are neither anxious nor stressed. Identifying the difference between being stressed or anxious and being worried (without any physical triggers) is important in helping us tackle the problem accordingly.

Stress and anxiety have similar symptoms, which means it can sometimes be hard to tell them apart. It's important to learn to identify stress and anxiety by being more aware of what we're feeling rather than putting a blanket statement of 'I'm stressed' over all our feelings and emotions; only then can we respond accordingly and support ourselves more effectively.

Three ways to identify whether you're stressed or anxious

1. **Anxiety is excessive.** Some situations are stressful and would be for anyone. However, if that feeling of worry is

unusual and excessive, or you find yourself catastrophizing, it may be anxiety rather than stress.

2. **Stress is usually external while anxiety is internal.** Stress is usually triggered by something external, like your boss reprimanding you, having an exam or juggling multiple responsibilities. If you were to remove the obvious stressors, would that feeling of overwhelm disappear? If yes, then it's stress. If no, then you're anxious.

3. **Anxiety causes feelings of fear over things that haven't happened yet.** Anxiety can be triggered without an external stressor. You may feel dread and apprehension when there isn't anything obvious to be concerned about. Stress, on the other hand, is an external pressure that's causing you to respond accordingly.

Anxiety often becomes ingrained in a habit loop, a neurological pattern involving cue, routine and reward. The cue might be a stressful trigger, initiating a habitual response to alleviate the discomfort. Individuals who suffer from anxiety may resort to repetitive behaviours or thought patterns as routine, attempting to regain a sense of control to mitigate perceived threats. The reward relief, though temporary, reinforces the loop, strengthening the connection between anxiety and the habitual response. Over time this cycle can become automatic, intensifying anxiety and making it challenging to break free from the loop. Understanding and disrupting this habit loop is crucial for managing anxiety and fostering healthier coping mechanisms.

Tools to break this habit loop[17]

Identify Triggers

Recognizing and understanding the cues that trigger the anxiety loop is the first step in interrupting the cycle.

Walking neuroscientifically reduces anxiety

As you go through the world, your surroundings move past your peripheral vision and the pattern of the environment flowing past your eyes causes you to look from side to side.[18] You're probably not aware of it and you're also not actively moving your eyes; it's something that happens naturally as you go about your day, provided you're not staring into the screen of your phone. This is called optokinetic nystagmus and it describes the rythmic side-to-side movements of the eyes as they track moving objects or scenery passing by. This lateral eye movement has been shown to deactivate the amygdala, the area in the brain known for its role in fear processing. That is because when your eyes are moving from side to side, they activate an area of the brain called the frontoparietal network, a collection of areas in the brain involved in attention, complex problem solving and working memory. This network competes for resources with the amygdala, thus switching the amygdala off and reducing anxiety.

As we've seen, anxiety and stress impair your executive functioning (part of your frontoparietal cortex), and when you're stressed you find it hard to concentrate on problem solving and make decisions, and your judgement is impaired. The opposite happens when you engage the frontoparietal network by moving your eyes from side to side and thus quieting the amygdala. One example of this is eye movement desensitization and reprocessing (EMDR) therapy, where a therapist uses goal-directed eye movements to help a patient process events and emotions without the fear aspect.

Another primal aspect of walking is that when you're in an open space and in nature, your brain can see that there is no imminent threat and thus reduces amygdala activity. Our brain is wired for survival, and research[19] shows that the amygdala is more active in people who live in cities compared to those who live in rural areas. Our brain is constantly reviewing the information that comes in and working out whether it's dangerous or not, and being in the city means your brain has more to process with all the moving cars, people and noises, etc. The research[20] also shows that the activity of the amygdala decreases as a result of just one hour of walking in nature; therefore being outside, in nature and in open spaces, can have beneficial anxiety-reducing effects on the brain.

When you're feeling stressed and anxious, one of the best tools is to go outside for a walk so that you can process your thoughts and emotions without the fear processing. This is where the phrase 'going for a walk to clear my head' comes from.

More tools for anxiety

- Anxiety requires some cognitive interventions to help people with how they perceive stressful events.
- Meditation has been shown to shrink the size of the amygdala.
- Self-hypnosis can help with the visualization of how one would feel in certain situations. Often our brain doesn't know how we should feel in particular scenarios so it reverts to what it knows best. Visualization can help people learn what their brain and body should feel in certain scenarios.

Neurochemicals and emotions

I've been reading Penguin books since I was a little girl, and in June 2023 I was due to meet my editor at the Penguin Random House head office in London. As I approached the doors, there I saw it, the iconic Penguin sign. I felt like I was living my best movie-scene moment. As I approached the front of house staff, I was smiling from ear to ear and I could barely believe this was happening. I confidently asked for my editor, my head held high, and I was so proud. Wow, so many emotions were running through my body.

A few days after this I threw a big birthday party and it was a total blast! There were performers and I even sang for my guests. It is an understatement to say I was elated and driven by pure happiness and joy. I felt euphoric, high on life. Happiness chemicals were rushing through my veins, propelling my inner social butterfly.

I landed back home in Portugal to my quiet rural life two days after the party, five days after my epic movie-scene moment at Penguin headquarters, knowing that my editor and I were a match made in writing heaven, yet I felt depressed. I was irritable and sad. I should've been happy. I should've been grateful for the incredible week I had just had in London celebrating all these wonderful moments, moments I'll remember for the rest of my life. But instead I was agitated and my mood was low.

You see, the thing is, with every high there comes a low. I was on an emotional comedown from the surge of happiness and gratitude I had felt the week before. Sometimes when we experience these extreme highs there is a moment of rebound whereby neurochemicals need to level out and replenish. Often people attribute their mood swings, irritability and depressive symptoms to something being wrong with them when usually it's the body's way of trying to reach homeostasis. Homeostasis is the body's way of self-regulating by readjusting internal states after they've been affected by an internal or external force. For example, when it's hot outside, your body starts to sweat to manage your internal temperature so that you don't spontaneously combust from the heat. My body was readjusting from the high I had experienced the week before. It doesn't mean I am an ungrateful person or that I am depressed, even though those were the feelings I was experiencing. It means that my brain was recovering from the emotional high of the week before.

A few weeks later, after losing the 2023 Wimbledon men's final to Carlos Alcaraz, Novak Djokovic said:

> *'When all the emotions are settled, I have to still be very grateful'*

When the presenter interviewing him after the match went on to say that he should still be proud of what he'd achieved, he replied, 'I will be, tomorrow morning, but for now it's a tough one to swallow.' And that for me is perfect. He is completely aware of his emotions, and he's aware that right now he's going to feel a drop from the loss he just experienced against Alcaraz, but tomorrow he will find gratitude and he'll be proud of all the other wins he's achieved. Djokovic is fascinating to me because of his mindset and the way he approaches winning and losing. When you think about it, all these amazing tennis players have great serves and swings. They know how to play through repetition and practice, but the thing that sets the top ten players in the world apart from the rest is their mindset.

Post-Olympic depression, a term coined by sports psychologist Scott Goldman, is very real. It describes the emotional drop that athletes feel after an Olympic event, both winners and losers. Physiological and psychological resources are both depleted, which can lead athletes to feel very low;[21] they are quite literally on an emotional comedown from participating in such a grand event. Many other athletes have reported the same feelings post-competition. This drop can lead to feelings of loss of purpose, low self-worth, confusion and severe depression. Post-Olympic depression can last weeks and it can become serious if left untreated.

The reason I'm highlighting the importance of these athletes' experiences with their neurochemicals is that sometimes we mere mortals are driven by the abundance or scarcity of these chemicals on a given day. They are temporary and dynamic, meaning they can fluctuate, yet they can heavily dictate how we perceive ourselves. This can keep us in a cycle of rumination and overthinking. Just by understanding this and our neurobiology, we can start to understand ourselves a bit better and learn to break the cycle. For all we know, Djokovic may have been keeping it together for the cameras, but from what I have seen in other interviews he acknowledges his losses and honours his feelings. His mindset towards the game is what allows him to use the losses as fuel and learning.

Being emotionally dysregulated means that we can experience heightened levels of stress, anxiety and even panic in response to mild stressors. This keeps us in a constant state of arousal, which can mimic a physiological response associated with being burned out. Additionally, emotional dysregulation can lead to a heightened perception of threats in the environment, even when those threats are not objectively present, which can cause hypervigilance and the anticipation of danger. And lastly, emotional dysregulation can have physical consequences, such as disruptions in sleep, digestion and immune function, which can further contribute to a state of heightened vigilance and heightened stress. We know that chronic stress can hinder us from creating meaningful plastic changes, hence why it's important to regulate our emotions so that we can continue to rewire our narrative.

How to manage this

When you find yourself in this state, I want you to understand that it is temporary. Like a wave, it always passes.

By understanding our brain and the neurochemicals that come into play when we're feeling like this, we gain a sense of control over the situation. This allows us to ride the wave until the moment passes.

In this scenario, creating a narrative for the situation can be helpful. We can do this by talking to friends and family about how we're feeling or by journalling.

Mental currency

How easy is it to waste ten minutes on social media without even realizing? My MSc research investigated how we allocate mental energy resources to social media, something that people usually engage in with the perceived idea that they are taking a mental break from work. Our brains have limited capacity for daily energy expenditure and therefore have limited energy resources.[22] Activities such as studying, problem solving and higher-order thinking can lead to increased energy consumption, thus leading to mental fatigue. The problem is that when we use social media during our work breaks as a form of distraction, thinking that we're giving our brains a break, our energy resources are, in fact, still being allocated to something that is mentally taxing – getting a kick from doom-scrolling. This leads to cognitive overload, a state where the mind is overwhelmed by excessive information or stimuli, hindering our brain processing. In today's digital age, constant data influx from sources like social media can lead to mental fatigue, impacting decision-making and overall cognitive function.

Social media platforms are designed to capture our attention, contributing significantly to this overload. The incessant notifications, updates and curated content demand our mental resources. So when you're on your lunch break and you're scrolling on social media, you are essentially giving away your mental currency to the online world. Then when you get back to your desk, whether you're a writer, a student, an office-goer, an architect or anyone who needs to use mental energy for their work, you have even more depleted energy resources – you haven't actually taken a mental break.

Imagine you are on a treadmill for eight hours a day (like the length of your working day), running at a steady pace all day long, and then in your lunch break you move on to the stationary bike (social media) and eat your lunch on that – you're not actually taking a break. This is why taking intentional mental breaks is essential. Stepping away from screens, engaging in mindfulness or pursuing offline activities allows the mind to recharge. You're probably thinking, *Well, what am I supposed to do?* But the thing is, you only need to take a proper strategic break for ten to fifteen minutes to gain the benefits of brain energy renewal. My research specifically looked at mindfulness meditation, and results showed that it can replenish mental energy resources. When participants took a strategic fifteen-minute break, their attention and the quality of the decisions they made improved. I call this a strategic break because it requires you to be intentional with how you spend your break. In an ideal world you would spend the first ten to fifteen minutes of your break with your eyes closed, switching off from the world, and then have your social media and coffee after. If you think about it, external stimuli constantly influence our state of mind, but how often during the day do we go 'inwards' and check in with ourselves? If you don't want to meditate, then all the strategic break requires you to do is to switch off from any external influence that can affect your system and spend ten to fifteen minutes replenishing your energy resources so you don't drive your brain to depletion. If you don't, then what tends to happen is you run on fumes for the rest of the day, and when you get home you're completely wasted and irritable and you take it out on those around you or make poor decisions. Or, worse, your brain takes a break for you when you least expect it.

In short, it's in our best interest to spend less time on social media and more time with our head in the real world, taking time out for ourselves during the day. Strategic breaks can help increase our attention. Over time our brain's energy resources become depleted, which we call vigilance decrement, meaning that our attention slowly decreases.

I see it with my dogs all the time. We'll be doing some training exercises, and then after ten minutes they become distracted, lose focus on the task and start making mistakes. I can translate that to my own work. It's easier for me to get into a flow of writing when I am 'fresh' or when I've taken a proper break. I avoid looking at my phone during my lunch breaks, and sometimes I will even close my eyes and 'nap' (I don't fall asleep) to reallocate energy to my brain.

We can increase that window of focus over time, provided that we're allowing the brain sufficient time to recover efficiently on a regular basis. The issue with vigilance decrement is that through mental heuristics (or mental shortcuts) when you're tired, you'll revert to what is automatic and deeply ingrained. Before you know it, you're back to repeating those negative behaviours, stress eating, doom-scrolling and hindering your mental capacity. Your brain will look for ways to take shortcuts, so that it doesn't have to think about the changes you're trying to make, and it'll keep reinforcing this pattern. The brain will find ways to preserve energy and take the route of least resistance if it can, and if you're tired and mentally drained, this is going to be amplified, because why would your brain prioritize what is less known if it can go down an easier route to preserve energy for more mentally demanding tasks?

Additionally, the more tired you become, the more you start relying on your phone for a dopamine boost. This is natural but all the scientific evidence points towards putting effort in for the reward. Social media is just too easy, which dysregulates our motivation drive and burns us out even more. When we spend a lot of time on social media, we spike our dopamine levels to a point where we need more dopamine to drive us to want to do anything else. This is a slippery slope because we lull ourselves into believing that the answer is more dopamine from social media, and so we feel motivated temporarily but the feeling quickly dissipates.

The more frequently we spike our dopamine, the more we need to make ourselves feel better but paradoxically the harder this becomes. We start attributing our laziness, lack of motivation and procrastination to a personality type like there's something wrong with us. We may have been labelled lazy and unmotivated by our peers or even ourselves. This is something that could be fixed by limiting social media use and other quick thrills. The literature on dopamine is vast and complex, but the underlying theme seems to be that reward needs to be attached to effort. If we experience dopamine rewards too often, we lower its potency. When we release dopamine from activities that require effort and work, we see a very different curvature in the dopamine release, meaning that we don't tend to yo-yo between highs and lows as drastically. Exercise, breathwork and achieving personal- or work-related goals, for example, release dopamine in ways that are sustainable throughout the day. This does not impact our motivation negatively but, in fact, could help us achieve it more easily.

Using your phone less might feel hard at first, but eventually it'll get easier as the habit starts to weaken. I almost can't wait for my break time because I can feel my brain needing a rest, and once it's solidified as a habit, you'll be so glad you took it on. It's one of the lowest-hanging fruits of brain health and one of the best ways to maintain brain energy levels throughout the day so that we don't come home after a long day and lose all control.

Closing your eyes and resting from the world is an invaluable tool that doesn't need to be too complicated; simply switching off from the external world can have incredible benefits for your brain capacity.[23]

REWIRING ESSENTIALS – BREAK THE CYCLE

- Stress is not always bad.
  - Alert stress is responsible for keeping you alert during tasks.
  - Acute stress is a short-term burst of energy that helps you find a solution to a problem, and helps get you out of it.
- Learning to recover from acute stress regularly can promote mental resilience.
- Chronic stress occurs when the demands placed on us exceed our ability to cope effectively – this is bad.
- Chronic stress can push us into living in low-power mode and impact our ability to think clearly by disrupting how we respond to stressors and making us more hypervigilant.
- Tools for stress management.
  - The physiological sigh – page 41
  - Hobbies – page 45
  - Tools for chronic stress – page 48
- Learning to differentiate between stress, anxiety and worry can empower us to manage each emotion better.
- Accurately naming your emotions can help reactivate the prefrontal cortex, which is responsible for more logical thinking – see page 59 for the Wheel of Emotions.
- Mental currency is where you spend your cognitive energy – be careful of how you spend it.

TYPES OF STRATEGIC BREAKS:
PRAY
SING
WALK
DANCE
STRETCH
MEDITATE
BREATHWORK
CLOSE YOUR EYES AND REST
SELF-HYPNOSIS (THERE ARE APPS FOR THIS)

WHEN TO IMPLEMENT THESE BREAKS:
WHEN YOU'RE TIRED
IN BETWEEN MEETINGS
IN YOUR LUNCH BREAK
AT THE END OF YOUR DAY
BEFORE AN IMPORTANT CALL
BEFORE AN IMPORTANT PRESENTATION
AFTER A STRESSFUL EVENT

YOUR BRAIN IS
PROGRAMMED TO FOCUS
ON THE NEGATIVE

BUT YOU CAN CHANGE IT

Negativity Bias

Neuroscience shows that negative emotions elicit a much larger response in the brain than positive ones, meaning that we tend to pay more attention to bad things and overlook the good ones, probably as a result of evolution.

I asked my followers what negative assumptions they make about themselves. Some of the responses actually made me well up.

If I face a challenge at work, I am going to fail and my
team will hate my proposals.
My crush won't like me for me.
I'm afraid to think something positive and get my hopes
up just to be let down.
I will never be the best, so why even try?
After having a normal work conversation with
a co-worker, I always walk away thinking I said
something silly.
It won't work out for me.
I'm a fat tub of lard.
I'm fat and ugly.
I am not unique.

People only invite me out to be polite; they don't really
want me there.
I am not worthy of what I want.
Whenever my partner is in a bad mood, I assume
they're cheating.
Less attention and communication mean they've
stopped liking me.
I will never be healed enough to truly love myself.
I'm dysfunctional and weird.
When I see people affording things I can't, I feel bad
about myself and think I can never do that.
They cancelled on me so it must be because
they don't like me.

These negative biases can show up as stories we tell ourselves; they can also show up as a behaviour. For example, that flatmate who always comes home to recount the malevolent and unfortunate events of the day in a melodramatic and embellishing tone: 'and then the train was four minutes late, which pissed me off because I was already late, and then the guy on the train did this', 'and then my boss said that', 'and then the whole day just kept getting worse and worse'.

Sometimes I like to plant this seed in social settings; I certainly do it with myself and my clients. I'll ask them, 'What were some of the small wins and big wins of your week?'

They usually start out by saying, 'Not that much,' and they'll continue with gloom and doom, 'My week was mostly bad,' and then as we start talking I start to point out, 'Ah! That's a small win.'

'Oh, yeah . . . I guess it is.'

And what's interesting is that I can see them slowly starting to realize that they had more wins than they had paid attention to. By the end of the activity I can see them gleaming with a smile on their face, talking about all the wonderful things that happened. Of course we don't ignore the negative but the light that shines on their perception of them is brighter, meaning that they tell me about the negative aspects of their week with a more neutral tone governed by less emotion and bias.

Just like the universe wasn't conspiring against me,
perhaps it isn't conspiring against you either.

I love asking my friends this too: 'What was your favourite part of that day/week/holiday?'

They say that people never forget how you made them feel, and by highlighting their wins of the week, we're guaranteeing a life of happy memories and happy feelings with one another, even if we stop being friends with them.

What's going on in your brain

We all have the tendency to jump to negative assumptions and dwell on the negative. We call this negativity bias. To return to the Wheel of Emotions we looked at on page 59, you may have noticed that out of the eight basic emotions, only two are associated with positive emotions and two are both negative and positive depending on the context.

- ✗ Anger
- ✗ Disgust
- ✗ Sadness
- ✗ Fear
- ☑ Trust
- ☑ Joy
- ● Surprise
- ● Anticipation

Have you ever noticed that if you're having a great day but then something bad happens, it tends to ruin your whole day? The science of this negativity bias concludes that when there are equal measures of good and bad emotions, the psychological effects of the bad ones outweigh the good ones. It has also been shown that negative stimuli in the brain carry more informational value than positive stimuli, which also requires more attention and cognitive processing. This also affects judgement and decision-making. When we make decisions, we put more weight on the negative aspects of a decision than the positive.

Neuroscience studies[24] have even recorded the magnitude of brain responses (event-related potentials) as a direct result of a specific sensory event. In one case a set of pictures was shown to participants, who were presented with neutral pictures as a control measure and then a mix of positive and negative ones.[25] The negative images elicited a much greater response in the brain than the positive ones, despite the fact that both the positive and negative pictures were equally stimulating. This shows how we tend to give more weight and attention to negative experiences and emotions. And not only this but the brain registers negative stimuli more easily than positive events too. In other words, we are more likely to focus on, remember and be affected by the negative events or aspects of a situation, even when positive elements are present as well.

This can show up in our lives in various ways. For example, if you receive a bunch of great comments on your social media post but one negative one, you're more likely to dwell on the negative comment for the rest of your day or week. This happens as a product of evolution. As a species we have had to be highly attuned to potential threats and dangers, which was crucial for survival. We saw that of the eight basic emotions the majority of them were negative. But we can change this bias. Just like Djokovic chose gratitude for his previous wins during his post-match interview about his loss to Alcaraz.

I hear a lot of people say that they're a very negative person and that their immediate reaction to situations is to respond with bitterness either by being self-critical and comparing themselves to others or by being critical of others. You hear your friend got a promotion and the first thing you think is that they don't deserve it, and you're shocked that they should be in that position. Or you think it's unfair because you know how they are in their personal life. But your bias is probably over-looking all the positive things about them, and why they do deserve that promotion. Another example is when you see a picture of your partner's ex, and you start looking for all the reasons why you're better or better-looking than them. Our brains like to focus on the negative, and if this sounds like you, you're not alone.

The most detrimental part about this negativity bias is that we all have a narrative that we repeat, a story that we tell ourselves and others about ourselves. For some that repetitive story is good and it works in their favour, but for others it's a narrative that is perpetuated through the lens of negativity and self-criticism. Perhaps you gained that narrative from your parents, peers and socioeconomic status. Perhaps you learned this narrative by observing the way your parents operated in the world and it has heavily hindered how you navigate your life now. For example, your mother was highly critical of her body and now you are too. Now you carry self-limiting beliefs about yourself that hold you back from being who you want to be, from achieving all the things you want to, because you don't believe in yourself.

But the brain can change and so can these beliefs.

UNBECOMING (aka long-term depression – the weakening of synapses)

Neurons that fail to sync, fail to link.

Acknowledging our brain's negativity bias is a great step towards understanding why we dwell on these events, so we can work on how not to do this going forward.

Ditch the Negative

Provided that you are not in a fight-or-flight state (high stress), focusing on the positive helps to reframe your view on a situation, shifting your mindset, which will eventually lead to you being less affected by negative events over time. Remember what we learned about self-regulation in the previous chapter: we know that we need to be in a calm state of mind to shift our mindset. If you're in a state of high arousal, you'll need to use one of the self-regulation tools first before reframing.

Use the physiological sigh to regulate your nervous system. As a recap:

- Take two short inhales (ideally through your nose, but if you can't, then through your mouth is fine).
- Pause for one second.
- Exhale long and slow through your mouth.

Remember that whatever thoughts and reactions are repeated are what will strengthen the pathways in your brain. So dwelling and ruminating on negative events will strengthen the neural pathways for negative thinking and worrying. However, we can absolutely change this default mode of thinking so that our automatic reaction is more indicative of a positive mindset.

AN ACTIONABLE EXAMPLE

You experience something that triggers a negative response.
You have a work-related discussion with your colleagues that leaves you feeling like you said something stupid.

Self-regulate (physiological sigh if needed).

Reframe.
Your insecurity means you're more likely to be nervous around other people, but you know that you didn't say anything stupid because this is your profession and area of expertise; you're just worried that you did.

Over time your automatic response will no longer be to jump to negative conclusions because those connections have been weakened.

Negativity bias

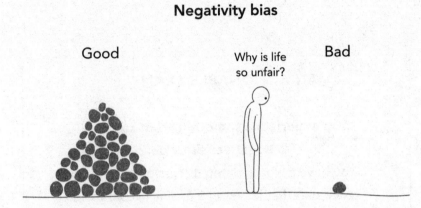

BECOMING (aka long-term potentiation – the strengthening of connections)

Neurons that fire together, wire together.

Dr Donald Hebb, the father of neuropsychology, was one of the first scientists to understand how the function of neurons contributes to psychological processes such as learning. He also discovered that when you repeat an experience, thought or action over and over, the brain learns to trigger the same neurons each time. In other words, if you repeat something, over time the same neurons are triggered to give you the same experience. And the more you repeat something, the stronger the connections become. In neuroscience we call this long-term potentiation – the strengthening of con-nections.

But neurons that fire apart, wire apart.

Yes indeed. We can unlearn certain behaviours. We can sever the co-firing of neurons so that they no longer activate one another. Conversely we call this long-term depression, which is not related to major depression. It's possible to change your automatic response to something. Hebbian laws explain that if neurons don't excite one another or communicate, then the connection of those neurons will weaken over time. So if you get triggered by something, and you leave enough time between the stimulus and the response, eventually those thoughts or neurons will no longer trigger one another and you won't react with an immediate response.

Let's look at this with another example. If you always feel the need to smoke a cigarette when you get stressed, that's because your brain has associated that solution as a relief for stress. The neurons in your brain are co-firing to say that the solution for stress is to smoke. It's an automatic reaction that most people don't pay attention to. But knowing that we can change this response means that we can substitute it with an alternative, so that over time your brain will take a different route when stressed instead of automatically thinking of smoking. It works in the same way with your thoughts. One might say that chang-ing thoughts is easier than quitting smoking because cigarettes have addictive properties, but often your brain jumps to nega-tivity because that's what has been practised over and over.

The great thing about plasticity is that it works both ways.

We can create new connections, but we can also undo some of them; we can unlearn some of the narratives that we repeat and replace them with more positive ones. You can sever two co-firing neurons by consciously inhibiting one thought from directly following the other. The more you delay a thought from following a precursor, the weaker the neural connections get. This knowledge will be important when we come to look in more detail at how to dismantle particular firing patterns.

The first law of neuroplasticity

I was making coffee this morning, as I have done for years. My partner recently bought this nifty little frother. It looks like a mixing stick with a halo on top that vibrates and spins so you can froth up the milk so it's velvety and, well, frothy. I've known for about a week now that you're only supposed to put a small amount of water or milk in the cup to froth it up before you top it up with the remaining water or milk. I was getting ready for the kettle to finish boiling so that I could put a tiny bit of water in the mug and get this magical halo frothing miracle out and make the best coffee ever ... but to my automatic detriment, even while I was thinking about it, I still managed to fill up the cup to the top with boiling water. Bugger! I knew not to do this and yet I did it anyway. Can you relate to this?

The first law of neuroplasticity states[26] that the repeated activation of a presynaptic neuron, one that is sending the signals, and a postsynaptic neuron, one that is receiving the signals, leads to an increase in the efficiency or effectiveness of communication between them. My inability to only put a small amount of water in the cup is the product of my neurons firing repeatedly in a particular manner, which means I operate without thought, with automaticity. The repeated activation of co-firing neurons leads to an increase in the effectiveness of communication between them. I was so used to making coffee in that particular way that subconscious patterns took over despite consciously telling myself that I should do it differently. Why am I telling you this? Well, this phenomenon appears in all aspects of life when it comes to our behaviour and how we operate. Most of the time we're going through life in automatic mode without really thinking, and negative thoughts can arise like they always have, and because those pathways have been strengthened through repetition we go down that well-trodden path . . . yet again.

The first law of neuroplasticity also states that the order and timing of the spikes (neurons communicating) determine the size and magnitude of a connection being strengthened (or weakened). I've been making coffee in a particular way, in a particular order, for probably around twelve years: that's twelve years of repeatedly filling the cup to the top with hot water. This connection is highly strengthened; it happens without thought. This is the same if you've been repeating negative stories to yourself for the same amount of time. I'm mentioning this because I want you to have compassion for yourself. We will change this together, but I really want you to understand why your brain does what it does. We need to reprogramme it and upgrade the software so that you can stop talking to yourself in a particular way. So that you stop jumping to negative conclusions, so that you stop seeing the negative in everything you do, so that you don't focus only on what you haven't achieved. I want you to see what's ahead of you on this journey of change. The problem with going through change is that it's hard to trust the process, not knowing whether things will work out. I totally appreciate that, but this is why people like me are here to guide you with science-backed advice. If you are struggling to trust the process, then I hope you're starting to have trust in me and this book, despite the fear that you're not making any progress.

The information you need to help the penny drop

You know how you beat yourself up all the time hoping it'll instil a change in you? The textbook example is getting upset with yourself about the cookie you just ate. Or it's repeating all those negative things about yourself and just hoping that one day you'll change. Well, as humans, we don't tend to learn from bad news (unless the news is so devastating that it initiates a change driven by an extreme emotion – fear). For example, you're told that the next cookie you eat will kill you. We tend to only want to see and hear about things that fit into our current beliefs, so we take in information that we want to hear and we ignore the information we don't. When your friend tells you your car is contributing to climate change, it's unlikely you're going to go straight to the garage and trade it in for a more eco-friendly one.

If you're a smoker, has someone telling you it's bad for your health ever helped you stop? We're quite good at avoiding information that doesn't fit our beliefs or might make us feel bad. It might cause us to question something, but usually that change comes from within, not from hearing the bad news from someone else. This is why going around in circles and beating yourself up and thinking negatively about yourself isn't driving any change. Telling yourself you're not good enough isn't going to make you feel better. Calling yourself fat when you eat the cookie isn't stopping you from eating that cookie the next time.

Studies[27] around the motivation of behavioural change show us that as humans we tend to ignore negative warnings and respond better to positive ones. When researchers[28,29] looked at the stock market index values between 2006 and 2008, they noticed that when the stock market was high, people would log in to their accounts to check the values all the time, and when the stock market was low, there were fewer log-ins. People were burying their heads in the sand, afraid of the negative outcomes that looking at a crashing account would bring. The only time people started frantically checking their accounts was when the market crashed later in 2008. What that tells us is that unless something is really, really bad, we don't tend to respond to negative information and it doesn't drive us to want to change. Not enough for it to actually make a positive difference. That's why you've been stuck in this cycle for so long. We have this negativity bias, and we tend to dwell on the negative, yet it doesn't make a difference to our behaviour when we use it as a form of motivation to change.

Reframe your thoughts

CHANGE YOUR NARRATIVE

I am not good enough	———▸	*I can adapt and learn*
I am defined by my trauma	——▸	*I will heal*
I don't like the way I look	——▸	*This body is my home*
I have no energy	——▸	*I need to rest*
I am lonely	——▸	*I can connect with nature*
I can't	——▸	*I can*
Nobody likes me	——▸	*I like myself*
I am bored	——▸	*There is so much to learn*

Gratitude

As we've already established, the brain, with its incredible plasticity, possesses the ability to reorganize and adapt by forming new neural connections and modifying existing ones. From a neuroscience perspective, gratitude involves the interplay of various regions of the brain and neurotransmitters that contribute to the experience and expression of feeling grateful about your life and overall more positive when life throws you lemons. In many ways gratitude is a skill, but the great thing about it is that the more you practise it, the more it reinforces positive thoughts, emotions and behaviours, creating a positive feedback loop.

This feedback loop continues to strengthen as you maintain a consistent gratitude practice. Over time your brain begins to wire itself to default to a more positive perspective, making it easier for you to notice and appreciate the good in your life. Gratitude becomes a natural and habitual response, leading to a more fulfilling and contented life.

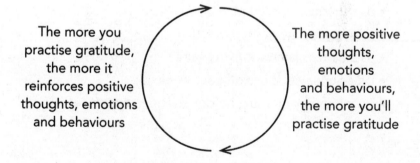

The more you practise gratitude, the more it reinforces positive thoughts, emotions and behaviours

The more positive thoughts, emotions and behaviours, the more you'll practise gratitude

Three ways to practise gratitude

Gratitude journal

Set aside a few minutes every day to write down things you're grateful for. It could be simple things like a beautiful sunset, a delicious meal, a friendly conversation or a personal accomplishment. Reflect on the day and jot down specific moments that brought you feelings of joy and gratitude.

Express gratitude to others

Take time to express your gratitude to the people around you. Tell a friend, family member or colleague how much you appreciate them and why. Handwritten notes, phone calls or face-to-face conversations can make a significant impact and strengthen your relationships.

Integrate a 'small wins and big wins of the week' into your practice

Incorporate this as an activity into your week. You can do it:

- as a journalling prompt
- during your meditation practice
- as a dinner game with your family
- on a Sunday to close off your week
- on a Monday morning before starting your week.

REWIRING ESSENTIALS – NEGATIVITY BIAS

- Negative emotions elicit a much larger response in the brain than positive ones.
- This means we have a tendency to pay more attention to bad things and overlook the good ones.
- Negative assumptions people make about themselves include:
    - If I face a challenge at work, I am going to fail and my team will hate my proposals.
    - My crush won't like me for me.
    - I'm afraid to think something positive and get my hopes up just to be let down.
    - It won't work out for me.
    - I am not unique.
- By pointing out the small wins in our day, we start to shift our narrative to see that there is some positivity among the bad.
- The most detrimental part about the negativity bias is that it skews the narrative we tell ourselves about ourselves to be worse than it is.
- Through neuroplasticity, we can learn to change our automatic response to be less negative, and change our narrative to tell a better story about ourselves.
- Reframing your thoughts helps.
- Finding gratitude helps rewire the brain to see things in a more positive light because it helps release chemicals associated with feeling good and happy.

- Practising gratitude is a self-fulfilling cycle – the more you practise it, the more it reinforces positive thoughts, and the more positive thoughts you have, the more you practise gratitude.

THE NEUROSCIENCE OF THE POWER OF BELIEF

STUDIES SHOW THAT OUR THOUGHTS CAN CREATE PATHWAYS IN OUR BRAINS TO STRENGTHEN OUR BELIEFS AND ULTIMATELY SHAPE OUR PERCEPTION

The Power of Your Thoughts

Have you ever heard the saying, 'If you knew how powerful your thoughts were, you'd never think a negative thought again?'

I want you to imagine that it's a hot summer day and you're making lemonade. You grab a handful of shiny lemons from the fridge; they're cold, bright yellow and waxy. You cut into one and the juice pours out of it. You begin to squeeze it, releasing a burst of zesty fragrance. Something compels you to take a big bite and your teeth wrap round the flesh, which is so juicy that it runs down your chin.

Did you salivate? If you did, then you just evoked a physiological mouth-watering response – as if it were right in front of you. This shows us how important our thoughts are. We can change our physiological response according to what we're thinking. So if you're perpetuating a negative narrative about yourself, you're reinforcing these pathways in your brain and strengthening these beliefs about yourself.

When I attend corporate speaking events, people usually gasp when I put into words how important our thoughts are.

Most people go through life with a background voice in their heads, constantly telling them a narrative, perpetuating a story about themselves. These stories are not always bad, but when they are, they push us down and reinforce our self-belief. We need to be careful of both the words we use in our heads and out in the world.

Josh, a client of mine, always jokes about how he doesn't make as much money as his wife. He jokes that it's only a matter of time before she realizes what a loser he is (his words, not mine) and that she'll leave him. I always remind him of how important the words he speaks are. Our words shape the way we think, and the way we think shapes how we feel; how we feel determines what we do, and what we do determines our belief. The words we use can have a downstream consequence that may seem minute when spoken, yet they trickle down into the core of our very being. If our words can shape the way we think, then they can alter our mood, shifting the neurochemicals in our brain to influence our emotions. Our words are so powerful yet we keep forgetting this . . . or perhaps we never realized how important they are in the first place.

When Josh jokes about those things, he's reinforcing an internal belief that somewhere deep down he must believe, or he wouldn't be joking about it in the first place. His words propel his actions, which manifest in the form of shame and a lack of trying to change. In many ways, it's easier to stay where we are because where we are, to our brain, is safe, despite it not being what's best for us.

Once I began explaining to Josh the importance of the words we use and their influence on our thoughts, he started to see a real shift in his behaviour. The less he perpetuates these negative narratives, the more self-esteem he's building. Our words can trap us behind walls, and it's only when we start breaking the pattern of behaviours that we can begin to realize we can be anyone we want to be outside of the box we've put ourselves in.

Mind-wandering thoughts are usually generated internally without an external stimulus. They use data from the past, present and expected future outcomes to generate an inner dialogue; they run in the background all the time, unstructured. These thoughts arise from a network of brain areas called the default mode network (DMN). And, just like the name suggests, it's your default mode of thinking. The DMN is a place of mind-wandering, creativity, internal thoughts and self-referential thinking. It's associated with introspective thoughts that run when you're not thinking about anything specific. It is often associated with autobiographical memories, social cognition and the processing of emotional states. However, it's also an area responsible for rumination. Studies[30] show that the DMN is often hyperactive in individuals who ruminate or repeat negative self-referential information. This hyperactivity can lead to a heightened focus on negative self-referential thoughts and an inability to shift attention away from them, perpetuating the cycle of rumination and contributing to the persistence of anxious thoughts and depressive symptoms.

What to do?

Metacognition

Metacognition can be a powerful tool for managing negative thoughts by enabling us to recognize, understand and regulate our thinking patterns.

AWARENESS OF NEGATIVE THINKING

By recognizing and acknowledging negative thoughts, we can begin to address them effectively.

REFRAMING

Once aware of negative thoughts, metacognition enables us to challenge and reframe them.

IDENTIFYING THINKING PATTERNS

Metacognition helps in identifying recurring negative thinking patterns. Understanding the root causes and triggers of negative thoughts allows us to proactively address them and develop healthier cognitive habits.

Meditation and nature

Meditation has been shown to deactivate the DMN,[31] disrupting the habitual cycle of negative thoughts and weakening the connections of ruminative thoughts. Meditation is like rebooting the brain's default mode of thinking, shifting from negative to more peaceful.

Spending time in nature also influences the DMN, promoting a state of relaxation and reducing the network's hyperactivity associated with self-referential thinking. This dual influence of meditation and nature on the DMN underscores their synergistic power in alleviating negative thinking and fostering mental well-being.

The only way is through

Initially, as you begin to meditate, it's not uncommon to become more aware of the constant stream of thoughts, which can feel overwhelming or even unsettling. However, with continued practice, you'll find that your relationship with your thoughts begins to change as you begin to sift through them.

Why you keep yourself busy

In tandem with the DMN there is the central executive network (CEN), which is responsible for external thought and more cognitively demanding tasks and problem solving. The CEN is like the CEO of your brain and it oversees the flow of information. It manages attention, decision-making and where we consciously invest our attention. These two networks work in tandem, meaning that when one network is more active, the other may be less active. For example, when you're engaged in an external task that requires your full attention, the CEN is more dominant and the DMN is suppressed. If you think about it, you wouldn't be pondering embarrassing teenage experiences while writing an email to your boss. I hope.

Think of the two networks like a see-saw.*,[32] When one is on, the other is off as the brain switches between internal and external processing. This is one of the reasons why people who struggle with ruminating thoughts and negative self-talk tend to keep themselves busy, which is often attributed to them being workaholics, because they want to escape their inner minds.

The salience network (SN) is the third major network and it is responsible for facilitating the switch between the CEN and DMN. It helps attach salience, meaning importance, to whatever is in your environment, and it helps to transition between these two networks based on the perceived importance of internal and external stimuli. For example, if you are working on your computer and hear a loud noise outside, the SN redirects your attention to that sound. When you are having a conversation with your friend, the SN is making your friend the most important person in your environment. The issue is that the SN can also be involved in directing attention towards ruminative thoughts and emotions. It can contribute to the persistent focus on these negative mental states. In the context of rumination,

* It's important to note that this is a general understanding of these networks and that there are instances when the two networks collaborate and interact. For example, when we engage in creative thinking there is often a combination of task-oriented processing gauged by the CEN as well as internal mind-wandering that fosters the creativity governed by the DMN. Effective problem solving also often requires switching between the two networks to find creative solutions. Thus while they are somewhat separate networks, they do overlap and interact with one another.

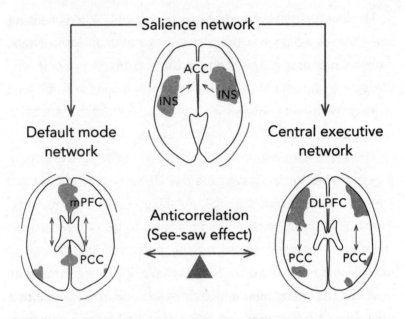

these interactions can lead to a heightened focus on negative self-referential thoughts and emotions, reinforcing the rumination cycle. Dysregulation of the SN can contribute to heightened rumination patterns by attaching importance to these thoughts. In conditions like depression and anxiety, there may be an overactivity of the SN in response to negative emotional stimuli, which can perpetuate rumination.

The solution?

Well, the thing with the DMN is that thoughts can run seemingly on automatic and they are also usually nonsensical, meaning that they are unstructured and free-flowing, with no real beginning or end. But we can force our thoughts into a more structured thought flow. For example, you're running your internal dialogue which highlights your negative self-talk, perpetuating thoughts of:

- I'm not good enough.
- I can't do this.
- Nobody likes me.
- I can't do anything right.
- It's always my fault.
- I take up too much space.

But as humans we have the ability to redirect our thoughts to something more coherent. We can drive our thoughts into a particular stream so that we can step away from the rumination and pay less attention to it.

Often rumination runs on repeat in the background because that's what the brain has practised. Remember that the brain is always going to revert to automatic and what it knows best, even if what it knows best is ruminating. By redirecting our thoughts and reframing them, we can start to create new pathways so that we can begin to dismantle the firing pattern of rumination. Eventually your automatic thoughts will change.

Step 1
Metacognition allows us to bring awareness to these thoughts and observe them.

Step 2
Understanding that negative emotional stimuli may have a bigger impact on us gives us a sense of autonomy. It's validating.

Step 3
Journalling or talking it out with a friend can help redirect your thoughts to something more cohesive. This is another example of how creating a narrative around your thoughts can be helpful. This will give your thoughts more structure so that they can stop running without beginning and end.

Emerging evidence shows that quieting negative thoughts and redirecting them away from the negative loop in your head can significantly reduce the memory of the thoughts and render them less vivid and anxiety-provoking.

As you can see in the picture below, when we think of words many areas of our brain are lit up. But when we direct them into something more coherent and generate them, we execute the frontal cortex, which is the more logical part of the brain. Earlier I explained that the frontal cortex is

Thinking of words   Generating words

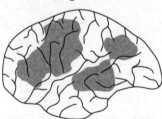 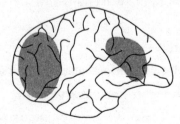

responsible for more logical thinking. This engages the part of your brain that is more reasonable, versus the limbic brain which thinks more emotionally. We also have to recruit other areas when speaking words, which means they are a lot less free-flowing and confusing, and more directed and cohesive.

I want to teach you how to tell people a story about yourself that is better than the one that you tell yourself, because the stories we tell ourselves and others can be very limiting. I want you to rewrite your narrative so that you can live to your utmost potential. Our thoughts are so powerful; they shape how we act and behave and carry ourselves. Our perception of ourselves directly influences how we present ourselves to the world, thus influencing how others perceive us too. So if we hold self-limiting beliefs or constantly speak to ourselves in a negative way, we're not only holding ourselves back but we're limiting others' opinions of us by extension. We all deserve a life where we walk around unapologetically, proud of who we are. I want you to get out of your own head and start exploring the world. When we're constantly worrying about our appearance or how people perceive us, or thinking negative things about ourselves, we lose the ability to really see the world around us and focus on expanding our minds.

One of the most important pieces of research on the power of our thoughts comes from a notable experiment[33] known as the piano experiment, conducted by Dr Alvaro Pascual-Leone. Participants were divided into three groups: a control group, a physical practice group and a mental practice group. The physical practice group was asked to learn a five-finger piece on the piano, and the mental practice group only had to *imagine* that they were learning that same five-finger piece, even though they never actually touched the keys of a piano. However, both groups had similar activity of brain plasticity, showing how the brain can create new changes in itself just by thinking about something. The mental practice group had created new pathways in the brain, through thought alone.

QUESTIONS TO ASK YOURSELF

- What are the narratives that you are repeating to yourself that hold you back from being the best version of yourself?
- What box have you put yourself into that means you live your life within the confines of those walls?

The problem with these narratives is that they are not just words in our head; they dictate a big part of our life and they bleed into everything we do.

THE WAY YOU SPEAK TO YOURSELF DICTATES:

YOUR MOOD

YOUR TRUTH

YOUR HABITS

YOUR SELF-WORTH

WHAT YOU ATTRACT

HOW PEOPLE SEE YOU

YOUR BODY LANGUAGE

WHO YOU WILL BECOME

HOW YOU GO ABOUT YOUR DAY

HOW YOU INFLUENCE THOSE AROUND YOU

THE WAY YOU SEE YOUR FRIENDS AND FAMILY

The famous milkshake experiment[34]

In 2011, Dr Alia Crum led an incredible research experiment based on mindsets and beliefs. She divided participants into two groups. One group was given a nutritional information sheet with a picture of a chocolate milkshake that described it as being rich and indulgent; it also told participants that it had 620 calories in it. The second group was told that they were drinking a sensible guilt-free milkshake of 140 calories. The nutrition labels were smart with how they depicted the milkshakes, leading participants to believe that they were either drinking a super-calorific drink or a fat- and sugar-free shake. What they observed in both groups was that their physiological response to hunger changed to the point where it actually altered their hormonal response. This means that their physiology changed in response to their beliefs.

The group who drank the indulgent shake had lower levels of ghrelin in their system an hour and a half after they drank the shake. Ghrelin is a hormone that signals to the brain whether you're hungry or not. Ghrelin levels rise when you haven't eaten in a while to signal that you should eat soon, and they drop after you've eaten a big meal to signal to the body that you're not hungry and you need to digest. So the indulgent group that exhibited lower levels of ghrelin would indicate higher levels of satiety and less hunger. On the other hand, the group that had the sensible low-calorie shake showed higher levels of ghrelin in their blood system an hour and a half after they drank the shake.

This means that their perceived idea of how calorific the shake was influenced how quickly they felt hungry after they drank it. I think you can guess what I'm about to say next ... and that is that, yes, both shakes had the same number of calories: 380.

Perception and belief can influence the way we experience the world around us

Participants in a study[35] were convinced, through hypnosis, to believe that their index fingers were five times smaller and then five times larger than their usual size. The researchers administered two pinpricks to their fingers; when they were told their fingers were five times larger, they felt the pinpricks as further apart than had been administered. And when they were told that their fingers were five times smaller, they couldn't distinguish between the two points; their ability to distinguish between the two points worsened as the perception of their hand size changed.

Our beliefs can change our perception of the world around us and how we interpret it.

I hope that this knowledge empowers you to see that you have power over your thoughts. Everything you've repeated up until now may be a fluke in your programming from childhood or external influence that has led you to reinforce a set of beliefs that may not necessarily be something you deeply believe or want for yourself. Your beliefs can change your perception of the world around you and how you interpret it, and as you move through this book, you'll start to accept a new reality for yourself. We're going to rewire your narrative.

REWIRING ESSENTIALS – THE POWER OF YOUR THOUGHTS

- Thoughts can create new pathways in your brain and that can strengthen or weaken a belief about yourself.
- If you perpetuate a negative narrative about yourself, you're reinforcing and strengthening these beliefs about yourself.
- To change this, we have something called metacognition – the ability to think and observe our thoughts.
- This can help us reframe our thoughts.
- Thoughts take up a lot of space in our minds, whereas generating words is harder because it uses a smaller part of our brain.
- Therefore, journalling is a great tool to help remove the emotional load from thoughts and beliefs.
- The way you speak to yourself dictates your mood, your habits, your body language and how you present yourself to the world.
- You can change this about yourself – it might be a learned behaviour that's ingrained, but we know that through neuroplasticity, we can weaken pathways of communication to undo these thought patterns.

DEATH BY A THOUSAND CUTS: CREEPING NORMALITY

A NEGATIVE CHANGE THAT IS ACCEPTED BECAUSE IT OCCURS SLOWLY

Creeping Normality

*Creeping normality: a negative and incremental
change that is accepted because it happens so slowly
that it goes by unnoticed.*

I want to validate something that I don't think we speak about enough, and that is creeping normality and microtraumas. I have met many people in my life who say things like, 'My child-hood wasn't that bad; I shouldn't feel this particular way,' but they grew up in a home where their parents constantly fought, their feelings were invalidated or they felt rejected by the people whose sole job was to protect them. Microtraumas are minor incidents that can seem insignificant in the moment and therefore go unaddressed or ignored. However, over time these incidents accumulate and have the potential to affect us negatively. These can also be experienced as someone con-stantly criticizing you or making subtle comments about your appearance.

The issue with this type of behaviour is that it's never really bad enough to reach a point where you think an intervention is needed or that it is worthy of change. These subtle and hurtful patterns can go unnoticed for so long, yet they leave a long-lasting effect on our self-worth, anxieties and stress. When we think of trauma we think about a terrible event or act of violence that changes us forever. But psychological trauma can also be caused by small and seemingly harmless events; when repeated, they have a cumulative effect on our psychological well-being and impact our way of thinking, leaving us with a vulnerability to mental illness, low self-worth and unhealthy coping mechanisms.

You don't necessarily have to have experienced a single traumatic event or to come from a broken household. Sometimes rejection from our absent parents can also keep us in a state of survival because our safety needs haven't been met. Sometimes it can be hearing our parents argue all the time, even if they're not shouting. Other times it can be the feeling of inadequacy when being compared to our siblings. I had a friend, Samantha, who grew up in a working-class family and whose parents loved each other very much. They barely fought and they were happy; they were not rich but they were able to go on holidays once a year. Their mortgage was low because her parents were smart with money from a young age. But she had a sister, Martha, who always did better than her at school, and she always felt like she was in competition with her and that her parents were always comparing them. Their parents also put them in boxes and said things to them like, 'Samantha is going to be a professional dancer; she's very athletic. And Martha is the smarter one, who is more academic.' Samantha, who was a dancer, also wanted to go to university and study architecture, but she always felt as though there was this expectation for her to become a professional dancer, even though she preferred to do it as a hobby. All these beliefs about who she is supposed to be were shaped early on by her family. They were not bad people, and they didn't know how they were shaping her future, but they kept emphasizing her role as the dancer and invalidated her feelings when she tried to speak up about it. They wanted a ballerina so badly that they forgot to even consider whether she was happy doing it, and, in turn, she shaped her life around believing she wasn't good enough to be anything else, that she wasn't smart enough to be an academic and that her only skill was being a good dancer. This took a huge toll on her confidence and her autonomy as a person; she felt forced to do something that she didn't want to do, and ultimately it kept her stuck.

I've had clients with similar concerns. One whose mother told her she wasn't very good at school and wasn't very smart. This client became a housewife and was even terrified of taking my group coaching course because she was scared that there would be an exam or that she would forget all the content; even worse, she was scared that people would do better than her at it, even though the programme is not designed in any way to show a progression or represent grades of any sort. She was happily married with multiple children and she loved her life, but she always felt that she couldn't do more; luckily for her she wasn't too upset by it, but it's crazy to think that her life stayed on course for the destination her mother had set for her, consciously or unconsciously. How many times have you done that? Gone down a particular route because of your subconscious beliefs about yourself?

Another example of perpetual behaviours that knock our confidence and keep us in a state of survival is the consistent emotional abuse we receive from a partner. They may not be physically abusive, so their behaviour stays under the radar, never quite bad enough for you to raise it as a concern, but they say mean things to you or about your appearance, they make you feel small and they invalidate your feelings by telling you that you're overreacting. This makes you feel weak and it can have an effect on your confidence; overall it can compromise your capacity for healthy relationships and trust. This kind of behaviour can be so subtle that you're left wondering if you're the crazy one for thinking that it's wrong because their actions never reach the threshold that allows you to flag it as dangerous. So it slowly seeps into your veins, like a drip that's feeding you toxic waste, and before you know it, you've been poisoned.

Creeping normality means that over time you start to believe that this is the new normal. This slow change is not perceived as negative and so your perception of life shifts without you realizing that you yourself have changed completely. One small cut doesn't hurt all that much, but a thousand cuts can kill you.

If this is you, you may have gained a lot of beliefs and written a narrative about yourself that doesn't necessarily align with who you really are, who you really want to be or who you know you could be. You formed these narratives through early experiences which could have had an impact on the trajectory of your life, reinforcing a particular belief. You may have formulated a negative narrative about yourself based on what people have said about you in the past. From lies, abuse and shame. This is not the real you. The story you tell is one that was pre-programmed by other people, but is it really yours? Who is it that you want to be? Because really, when you think about it, we can be whoever we want to be if we can break free from the automatic behaviours dictated by our history. Luckily, deep in a fold within the cerebral cortex, between the temporal and frontal lobes, humans have something called the insula, a pea-sized structure responsible for conscious desire. It's a part of the brain that is so small yet it can trigger a cascade of emotions that will either make you want to get up and change your life or leave you heartbroken while you wreck it. You're here because you have a desire to do something about the narrative you keep telling yourself. To change your mind and rewire your narrative to be whoever you want to be.

What is it that triggers this desire? What threshold is required to stimulate the insula so much that it'll make you want to get up and change your life? I can't tell you that, and I'm not sure anyone else knows either. A sensory stimulus that triggers a cascade of sequences? A song? A movie? Or random chaos, also known as probability. If you're religious, you may put it down to God, but, whatever it is, I hope it inspires you to become whoever you want to be. This book and the science that it highlights will help you rewire these narratives, so that you can start telling yourself and the world a better story.

You've changed   I'd hope so

**Things you may have perpetuated over the years,
which you've learned to accept as the new normal:**

I'M SHY.
I'M LAZY.
I CAN'T DO THAT.
I'M NOT WORTHY OF LOVE.
I'M NOT GOOD ENOUGH TO BE . . .

We need to reconsider these narratives by assessing our values and what we truly believe about ourselves.

ACTIVITY

What are your core values?
(Bravery, fun, generosity, integrity, loyalty, creativity, curiosity, optimism, etc.)

_____

_____

Ditch the Negative

What is your definition of an 'ideal person'?

_____

_____

What do you have in common with this person?

_____

_____

Do they share similar core values with you?

_____

_____

What are the biggest differences?

_____

_____

Now here's the thing: usually, when people do this activity, they start to realize that the person they described as their ideal person is some version of themselves, if not their actual selves. It's always been you. It's like something clicks, in that perhaps you're more aligned with your own values than they had previously thought. You've been living this preprogrammed life based on how you think you should be because of what others have told you. But what if you broke free from those beliefs and started living a life that aligns with who you really are and who you really want to be? It's always been you! You are your own ideal person.

So, forget what they said, forget the lies you've been perpetuating in your head – who are you to you? Who do you want to be?

You can, if you so wish, create yourself and be whoever you want to be.

REWIRING ESSENTIALS – CREEPING NORMALITY

- Creeping normality is a negative and incremental change that is accepted because it happens so slowly that it goes by unnoticed.
- Sometimes, we experience minor incidents that can seem insignificant in the moment and therefore go unaddressed or ignored.
- Over time these incidents accumulate and have the potential to affect us negatively.
- These can also be experienced as someone constantly criticizing you or making subtle comments about your appearance.
- These subtle changes can manifest themselves into self-beliefs, so you start to believe that you're shy, lazy or not good enough.
- Reconnect with your core values to figure out who you really are.
- It's always been you.

YOUR BRAIN LIKES TO BE CORRECT AND IT HAS A CONFIRMATION BIAS

IF YOU TELL YOURSELF YOU'RE HAVING A BAD DAY, YOU'LL GO THROUGH THE WHOLE DAY FINDING REASONS AS TO WHY YOUR DAY IS DOOMED.

THE SAME GOES FOR THE THINGS THAT WE REPEAT TO OURSELVES.

I AM NOT ATTRACTIVE | I AM NOT GOOD ENOUGH | I AM NOT INTELLIGENT | NOBODY LIKES ME

You'll See
It When You
Believe It

Caroline: 'Don't look right now but look at how cute that guy at the bar is.'

Lucia looks over in a not-so-inconspicuous manner and the cute guy sees her. Now the girls are giggling to each other right after looking at him . . . because he's cute.

Tom (cute guy) immediately thinks there's something wrong with him. 'Were they laughing at something on my face?' he asks his friend.

Tom was certain he was being picked at rather than admired. It's unfathomable to him that they could have been checking him out.

You know that feeling when you walk into a room and you think everyone is judging you or talking about you? The feeling when you leave a conversation with your co-workers, and think you said something silly and that they don't like you? You know that feeling when your partner is having an off day and you immediately think it has something to do with you? That's your belief driving the way you see things.

On the flip side, there are times when you suddenly snap out of it and see something from a different perspective. That ah-ha moment. The one where you realize you've been dating clowns your whole life despite your friends telling you countless times that they're not funny. You know? That realization when you come to your senses and you think, *What was I thinking?* Well, don't beat yourself up about it. As humans, we see things as we believe them to be. That's the wonderful thing about your brain; it helps you live in delulu land until you finally see something, and then you just can't unsee it.

What's going on in your brain

Our brain has stored templates of previous experiences, memories and scenarios. And when we're going about our daily lives, the brain is constantly checking new experiences with previously stored templates of similar situations so that it can decide whether it's worthy of being brought to your conscious attention. This is governed by the SN I mentioned earlier. If something is operating with the same level of automaticity as usual, then unless you pay attention to it by redirecting your thoughts and telling yourself that something is of importance, it'll just keep repeating. The purpose of these memory templates is to ensure that the brain is being energy-efficient and doesn't encode new memories about the situation you're in if your brain has experienced it before.

Think of it like this: if you had to be constantly checking your surroundings and making sure that everything around you was correct and that you were executing your actions in the same way every day, you would be very tired by noon. This memory-template system checks everything on a subconscious level and only brings it to your attention if needed. If you see things with a particular view or bias, you'll go through life with that template in mind and see things the way you believe them to be. So if your ex cheated on you, now you have a template for relationships which tells you that your future lover will too, so you start to micromanage the relationship and find proof that you're correct.

Let's look at these memory templates using another example. When you're walking on the pavement on your way to work, you're not paying attention to whether the other people walking on the pavement are doing the correct thing. You're also not checking all the other parts of your walk, like the trees, the bins, the birds and the models of cars you see. Sure, you could pay attention to all that if you wanted to, but you'd have to tell your brain that all these things are relevant. Chances are that if you're walking to work on your usual commute, you're just walking on autopilot and not paying that much attention to what is around you. Your brain is also not making new memories about the situation if everything you see on that walk is pretty standard compared to what you usually encounter on your morning commute. The pattern of neuronal activity that is associated with your walk is covered by previous memory templates that your brain has made, and nothing new is there to change that. Therefore no new memories are encoded, and your attention stays on getting to work, or your phone or the argument you had with your partner the previous night.

However, if you saw a big beach ball rolling down the centre of the road in the same flow of traffic, your brain would notice that because it is not associated with a common memory template in your brain. The SN would be activated and your brain would bring that information to your conscious brain and make you alert. *This is weird. Why is there a beach ball on a busy London road?* Equally, if the bins were out on a different day than they should be, or something grabbed your attention like a handsome great Dane, then you would snap out of your own internal thoughts, pay attention to your environment and make new memories about the event, especially if the beach ball started to cause accidents on the road. That would be a very unusual image, so higher levels of noradrenaline and acetylcholine would be present, and the plasticity of the memory would consolidate more quickly and easily because that image and scenario would be of utmost importance.

The reticular activating system

In the introduction I briefly introduced the RAS. We discussed the friend in the coffee shop scenario and how you could choose to pay attention to their words or the noise that's outside the coffee shop. You have the ability to decide what's relevant and shift your attention like a spotlight to whatever is more relevant in the moment. Have you ever experienced that when you want to buy something and then suddenly you see it everywhere? For example, you decide you want to buy a royal blue BMW car, and initially you felt like that's a unique colour for a car, but as soon as you decide you want one, you start seeing them everywhere. Just like the friend in the coffee shop or your boring morning commute, your brain filters out images and sounds from your environment until you attach importance to them.

It has been hypothesized that parents who live close to airports are able to sleep through a plane taking off in the night yet will wake up if their baby shuffles around in the next room. The RAS has learned through experience that the plane is not important to them, but that the baby potentially being awake is.

I remember staying with my sister in her flat in Hackney. The thing with London homes is that they all come with some kind of quirk; hers was that the smoke detector would beep twice every five minutes. The interesting thing is that when I mentioned to her how annoying the sound was, she didn't know what I was talking about. *What? I can't sleep properly through the night because of that annoying beep and she doesn't know what I'm talking about?* So I pointed out to her where the sound was coming from. 'Ah yes, that thing. I forgot about that.' Fascinating. She wakes up in a panic when her baby farts but she can't hear the regular beep right outside her bedroom door. But that's just it . . . it's a *regular* beep. The beep had been beeping for as long as she could remember and so her brain had learned to filter out the sound. Just like you've learned to filter out other information from your brain when you go about your day. This means you sort of live in 'your own bubble'. Your partner is not being unfaithful; you just can't see it any other way. But just like we can tell the brain what's important, we tend to do that with other behaviours too. For example, you have a bad morning and then you tell yourself you're having a bad day. The brain likes to be correct, so when we announce such things as 'today is the worst day ever' or 'I am not good enough', they become a self-fulfilling prophecy. You may not be doing this consciously but you are doing it. Just like you're looking for reasons to prove that your partner is being sketchy.

Confirmation bias

Sometimes we move through life governed by these memory templates and they form a confirmation bias. Confirmation bias is our tendency to interpret, search for and recall information to confirm our prior beliefs or values. So in the 'partner cheating' scenario, you may be constantly looking for cues to prove that your partner is more distant and less interested in you.

In the eighties, researchers led an interesting series of experiments that illustrated how our body image and bias affect how we perceive ourselves. Participants of one of the studies[36] were asked to go into a job interview with facial scars applied by a make-up artist. Unbeknown to them, the scar was removed before the interview; however, the participants still reported feeling discomfort in the social setting. The study highlighted the psychological impact of perceived flaws on self-esteem and behaviour. The participants reported feeling powerless, and showed higher levels of distress and low self-esteem. They also reported that the interviewer showed bias towards them and that they were staring at their scars and were ruder to them. These results highlight our perceived bias about what we think is happening around us. We see what we believe.

Most of the time our actions are unconscious, and it's only when we bring awareness to them that we may realize that they need a change. So if you enter a meeting room with low self-confidence, shoulders hunched down, then that is how you're going to enter the room every single time. If you repeat a particular belief to yourself, your brain will subconsciously match those beliefs with previous memory templates that can prove that theory. You'll go through life confirming that bias too because the brain wants to be correct.

If you tell yourself you're having a bad day, your brain will find ways to reinforce that belief and you'll go about the rest of your day finding ways to prove that this day is bad.

And so it is with negative self-beliefs. When you believe that you are not worthy, or not confident, or you have a negative belief about yourself, your body language follows that belief. Moreover, the brain perceives your behaviour as normal and stops paying conscious attention to it, and before you know it, you've snow-balled to further reinforce this belief with everything you do.

If you go through life believing that you are not attractive or smart, you will find ways to prove yourself right.

Interestingly, research[37] shows that people are liked more than they know, despite people believing that they were awkward in social conversations and had perceived being unliked. Shifting our narrative could help us see things in a different light. That ah-ha moment I spoke of. Again, perhaps the world (and the people in it) is not conspiring against you.

The cognitive triangle

The cognitive triangle forms the basis of cognitive behavioural therapy (CBT), a therapy technique that helps manage your problems by changing the way you think and behave. The cognitive triangle visually represents how our thoughts, emotions and behaviours impact one another. What we think affects how we feel, which ultimately affects how we respond to the situation.

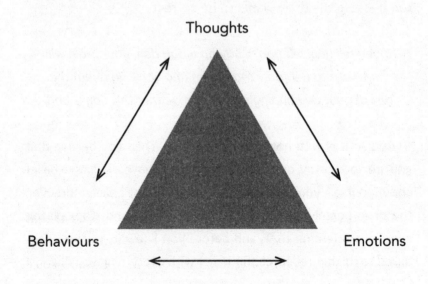

For example, you don't think you're good enough for the presentation you're about to give. You start feeling insecure, nervous and anxious about the situation. Your body language follows this and so your presentation is weak. This can further feed in to your confirmation bias because your peers will be able to see that you're not confident, so this has a knock-on effect on your belief and further reinforces it.

The laws of neuroplasticity show us how we can quickly learn a particular behaviour or association; if you keep repeating stories to yourself like 'I am not good enough', 'nobody likes me' or 'I'm a failure', then by means of plasticity you're reinforcing these beliefs. We've also seen how these beliefs and thoughts tie in with our behaviour, body language and the things we attract into our life, the way we perceive the world and the confirmation bias we go around trying to prove. These narratives are keeping you stuck, and they can dictate the majority of your life through the subconscious means of manifestation.

You'll see it when you believe it. When you believe something deep down, it shapes the way you see the world, the way you interact with it, how you behave towards the people around you. It shapes what you attract based on the body language you're showing the world. You'll see it when you believe it because what we believe shapes what we see. Have you ever considered that you are not a failure? That you are capable of the things you've been dreaming about?

What if you change the narrative and try to find proof that the world is, in fact, conspiring with you, not against you, to make you happy and help you get everything you want. You'll see it when you believe it: the world is not conspiring against you. You and your brain are conspiring against you by telling yourself lies and then looking for proof to reinforce these lies. It's time to get real with ourselves. It's time to bring all these things to the surface, to our attention, so that they stop running in the background on automatic without our awareness. We're taking control of the steering wheel. Changing these memory templates. Telling our brain we have something new and important to think about: this change and our old patterns.

In 'Negativity Bias' you identified these biases and negative beliefs and reflected on them. What lies have you been telling yourself to fit your narrative? To prove that you're not worthy of this beautiful life?

'Until you make the unconscious conscious, it will direct your life and you will call it fate' – Carl Jung

REWIRING ESSENTIALS – YOU'LL SEE IT WHEN YOU BELIEVE IT

- When you tell yourself that you're having a bad day, your brain will start looking for things to prove that's true.
- The reticular activating system in your brain is responsible for filtering out information that is important and what isn't.
- The words you speak and the beliefs you hold can cause you to look for ways to confirm you're right.
- So when you announce things such as 'today is the worst day ever' or 'I am not good enough', they become a self-fulfilling prophecy.
- Confirmation bias is our tendency to interpret, search for and recall information to confirm our prior beliefs or values.
- Your thoughts, emotions and behaviours impact one another. What you think affects how you feel, which ultimately affects how you respond to the situation.
- When you believe something deep down, it shapes the way you see the world, the way you interact in it and how you behave towards the people around you.
- Change the narrative and try to find proof that the world is, in fact, conspiring with you, not against you, to make you happy and help you get everything you want.

GRIEF: ENDINGS AND LOSSES

WE'VE ALL LOST SOMETHING. WHETHER IT'S SOMEONE, THROUGH DEATH OR A BREAK-UP, OR PERHAPS EVEN A DREAM. SOMETIMES A LOSS CAN ALSO BE MOURNING THE LIFE YOU WERE SUPPOSED TO HAVE.

LOSSES AND ENDINGS ARE A FORM OF GRIEF THAT CAN CAUSE A WIDE RANGE OF PROBLEMS INCLUDING INSOMNIA, DEPRESSION AND LOW SELF-WORTH.

Endings, Loss and Grief

Have you ever felt like you were having a heart attack and were going to die? Your heart aches so much that you didn't know if you'd survive the feeling of agony and heartbreak? This is called broken-heart syndrome; it's a temporary heart condition with symptoms like those of a heart attack. It comes on very suddenly, but you can recover from it quickly too.

I remember being curled up on the floor of the biomedicine lab at my university, terrified that I was dying. I had never heard of broken-heart syndrome before, but in that moment I knew that this was more than a panic attack. Thankfully I was taking a physiology class and the PhD student who was responsible for the first-year neuroscience students was doing his thesis on stress and the heart.

'You're not dying,' he said, as I kept asking whether this was it for me.

I remember thinking that I was too young to have a heart attack. Eventually the feeling subsided and I was assessed by a doctor, who told me about broken-heart syndrome, a condition that can cause rapid but reversible weakness of the heart. It can mimic the symptoms of a heart attack and is generally brought on by extreme stress. Minutes before this 'heart attack', we were measuring our resting heart rates as part of the lesson. Mine was consistently sitting at around 110 beats per minute and my lab friend jokingly said I was about to spontaneously combust. My usual resting heart rate is around 55 beats per minute.

You don't need to know much about cardiac health to know that 110 resting is high and not normal for a fit twenty-six-year-old. I had just moved to a new city to study neuroscience and my boyfriend had dumped me two days after arriving, when he was supposed to come with me. I was grieving. It was a lot for me to deal with: a new school, a new environment and returning to full-time education after taking eight years out. The relationship with my ex was extremely toxic and the break-up was traumatic. I wanted all my secrets back from him but I couldn't get that back and my heart temporarily gave up. I was in physical anguish, mourning the loss of the life I'd had before this new beginning. My life in London, my friends, my dreams and, of course, the potential dog I was going to get with my ex. Thankfully I lost the ex, even though at the time I was not thankful at all.

I started to think about all the loss I had experienced in my life: family members, opportunities, the fact that I didn't get into medical school. My whole life I thought I was going to go to medical school; I had been speaking about it since my father died when I was eleven and I had made him a promise I would go. I had applied countless times and every time I was hit with a blow – eleven blows to be exact. Remember my eleven rejections on your next bad day. But see here's the major realization I had lying in bed one night . . . I don't think that deep down I ever wanted to be a doctor. I can't be sure but I have an inkling that it was just a narrative I had repeated to myself for so long, and so I was more attached to the name and trajectory than the actual profession. Because if I had become one, I'm pretty sure I'd have ended up in psychiatry or neurology anyway. Basically still studying brains.

I thought about my best friend that I had fallen out with for good. I was in agony when we stopped being friends. We don't talk about friendship break-ups enough, but they're real. And when speaking to others about how *endings* can affect us, I came to learn that most people were grieving something. Some were grieving people; others were grieving lost opportunities. Grieving that they too didn't get to become what they always thought they would and go into the profession they always dreamed of. Endings seem to hurt for all of us, and depending on what it is you've lost, the magnitude of that pain can vary. Some are incomprehensible, like the loss of a loved one, and others are easier to navigate. Nonetheless, endings are painful.

Endings can cause extreme stress and grief that can lead to a range of issues, including sleep problems and depression. In my case they even led to the feeling of a heart attack. Studies[38] have shown that the brain reacts to the pain of a loss in the same way that it would to physical pain. There is even some research[39] to show similarities to the pain of breaking a bone.

We've all lost something, whether it's somebody, a dream or a job, or whether it's Djokovic losing at Wimbledon or even experiencing his loss for him. It feels like you're in hell, and we've all experienced hell in one form or another, just different devils on different levels.

Grief also means grieving the life and future you should've had.

From a neuroplastic perspective, when you lose someone your brain is having to readjust to a completely new way of living. Over time the person you love has been ingrained into the network of neurons in your brain. Your neurons have associated certain aspects of your life with that person. You leave work and your initial response is to call them. You hear good news and your first thought is to turn round and tell them. Your neuronal firing pattern is operating in a sequence that it knows well. So when we lose someone, through death, a break-up or a falling-out, it can be very painful because we no longer see, talk to or touch that person. Our brain's firing pattern is no longer operating the way it used to and that can be very taxing for the brain.

It is similar for other types of loss and grief. If you were preparing to realize a dream, such as studying for medical school, training to be a Marine or sacrificing your social life to get into a particular programme, but then failed at it, this is going to feel very painful.

Losing something or someone can cause insomnia, so you

stay up thinking and ruminating over what you could have done differently. Such loss can elevate your stress levels and keep you in a state of survival mode. They can increase your anxiety and even leave you with feelings of depression. If this is you right now, I know that it seems hard to believe, but I promise you will get through it. This chapter will equip you with the tools to navigate endings so that you can transition into new beginnings and shift your narrative more smoothly.

What's going on in your brain

On a neurological level this is what your brain is doing when you're experiencing a loss. The first thing to understand is that your pain is real. Studies[40] show that areas of the brain responsible for pain processing and modulation are active when we're experiencing a loss. This tells us that the pain we are feeling is not only mental but also physical. When we experience pain, an area of the brain called the periaqueductal gray is active; this area is responsible for releasing norepinephrine into the system to dampen the signals coming from the body so that it can provide analgesic relief. Norepinephrine is also responsible for alertness in the brain, which could potentially be contributing to your inability to sleep and making you feel exhausted by the pain. That's because your brain is working hard to dampen the signals and make things easier for you.

Other active areas include the posterior cingulate cortex,[41] which is responsible for autobiographical memories; this is interesting because one of the most common underlying themes with grief and loss is the going-over of memories in our head. Since this area of the brain is more dominant during the processing of loss and endings, we can start to understand why we keep replaying events in our heads over and over like a movie reel that is keeping us awake. Our memories become more vivid and active, which is why we tend to go down a rabbit hole of reminiscing and wondering how we could have done things differently. We tend to remember the smallest things about a person or event. This can be very painful but it can also be a cathartic experience to help us get through the pain. We should embrace the brain's ability to go down memory lane and find ways to cherish the moments that we may eventually forget.

The anterior cingulate is our error-detection area.[42] It monitors our performance over time by detecting errors in our life and tasks, to help us find a better solution for future problems. This means that your brain will be constantly trying to find a solution to the ending you've suffered, and perhaps you will ruminate on what should have been or could have been. By acquiring this knowledge we can start to reframe our behaviour as a useful tool for the future. Djokovic says that he uses his failures and past experiences to be better prepared for the future. When your brain uses previous experiences to shape your world, we call this experience-dependent plasticity. Having experienced a loss means we can be better prepared next time.

Hyperactivity of the amygdala is another key response to grief and this type of activity is associated with impaired emotional processing and depression. The amygdala is responsible for emotional processing and detecting fear and danger. This can also activate our stress response, resulting in our bodies being flooded with stress hormones, which can contribute to the emotional and psychological response associated with grief, such as sleep disturbances, changes in appetite and emotional arousal. Additionally, the emotional brain takes over logical thinking, which can make us act on our impulses. It compromises our ability to make decisions and therefore we might find ourselves having trouble concentrating and remembering things. To add to that, areas of the brain involved in motivation (the ventral tegmental area) and craving (the nucleus accumbens) light up.[43] So during a break-up this can make you want to act upon your impulses and perhaps message your ex, check up on them or, even worse, lie to yourself about why you should get back together with them.

Loss and endings hurt because we have lost something, but within it we have also lost our routines, habits, automatic reactions and behaviours. Wanting to call someone as soon as you hear good news. The coffee you would normally make for them in the morning. This applies when the loss is not person-related too. Perhaps your morning commute is different since you lost your job or you're out of sorts and out of routine because you're at home looking for jobs.

When we experience loss there's a huge shift in our neuronal firing; the brain uses memories to predict how we should go about our day, but the brain needs to adjust to a new reality and play catch-up with life's events. Our automaticity dictates that we should get up, make a coffee and go to work. And it can feel weird if we don't have that any more. The loss of our sense of comfort and familiarity puts us in a place of fear, a place that we need to re-explore while also trying to navigate the pain of it all. But over time the brain will start adapting and rewiring itself to accept a life without what was lost, and that's when you'll start feeling a sense of acceptance. Give yourself grace. Thank yourself for being human and being able to feel anything at all. I'm holding your hand through this; we'll get through it.

Neurochemical solutions to endings and loss

When we lose someone or something, our brain's feel-good neurotransmitters, such as oxytocin, dopamine, serotonin and endorphins, tend to drop. This can drive us to feel anxious, depressed and isolated. Our brain will find ways to try to re-place these neurochemicals, but how we choose to replenish our resources is vitally important for our healing journey. We can pursue an active and healthy coping strategy or we can prolong our misery by engaging in unhealthy behaviours that hold us back from getting clarity and relief.

Here we discuss some interventions that we can adopt to help us on our healing journey through loss and endings.

Dopamine

When we're experiencing an ending or a loss, activities that may have previously been enjoyable may be less appealing due to decreased activity in the brain's reward system. Depending on the severity of the loss, this can cause anhedonia, an inability to experience pleasure. Motivation and arousal may also be affected when experiencing endings and loss, which may leave us feeling unmotivated to engage in daily activities that we know may help us feel better, yet it feels impossible to get ourselves out of the door. It's important to understand that these feelings are temporary, and with appropriate therapeutic interventions we can make sense of the loss and eventually return to activities that bring us joy again.

Impaired dopamine activity may drive us to engage in impulsive behaviours such as online shopping, stalking an ex on social media or finding new pieces of information that contribute to the missing pieces of a puzzle you've been trying to solve. For example, you're trying to figure out how serious your ex's new relationship is by seeing if their new partner is going on holiday with their family. If you're going through a break-up, reward pathways associated with motivation to get pleasure will trick you into believing that you need to text or call that person for a dopamine hit. If you didn't get your dream job, you may start looking for ways to find out who they hired instead of you.

This is all understandable behaviour, and it's important to acknowledge that it's grounded in dopaminergic activity, in an attempt to gain information to make yourself feel better. Your brain is essentially going through a withdrawal period and it's finding ways to replace the neurochemicals. Because of this some people may turn to drugs, alcohol and other addictive behaviours to fulfil dopamine needs. Substances are an understandable choice because they also help us dissociate from our feelings and the real world, so that we can numb the pain and feel something other than our emotional anguish for a period of time. Until we sober up and the dopamine wears off and we remember that we're still in pain, which pushes us down to another layer of hell that is filled with depression and despair. The compounding effects of the substances wearing off coupled with grief means we feel more inclined to repeat the behaviour until it becomes a vicious cycle of trying to find relief fuelled by uncontrollable addiction.

While it may seem hard to motivate yourself to do something more positively rewarding, such as exercise or going out to see your friends, it would be wise to adopt more positive behaviour towards dopamine release that isn't online shopping and information-seeking activities such as the ones described above. Exercise helps release dopamine and can serve as an adaptive coping mechanism during grieving. It can help improve your mood and alleviate stress. It may seem impossible to engage in anything like this but there are many alternatives.

Start with what you feel comfortable with. You could begin by stretching at home or doing a video workout. Or perhaps you could take up a new hobby that will keep you fulfilled, like learning to play the piano or knit. Or perhaps you like to bake and cook at home. Or it may be something a bit more extreme, such as taking up martial arts or hiking. Whatever it is, I encourage you to do something that will improve your mood as well as your dopamine. I very much doubt that checking out your ex's new partner is making you feel good about yourself.

INCREASE DOPAMINE
ENGAGE IN REWARDING ACTIVITIES, EVEN
THOUGH YOU MAY LACK MOTIVATION BECAUSE
LOW DOPAMINE MAY DRIVE YOU TO ENGAGE IN
LESS POSITIVE BEHAVIOURS IN AN ATTEMPT TO
FIND RELIEF.

Serotonin

Serotonin is responsible for mood modulation and it helps prevent extreme mood swings. When we're experiencing a loss, disruption in serotonin levels can contribute to mood disturbances such as heightened sadness, but more importantly it can heighten obsessive thoughts, irritability and negative thinking. This can lead us down a pathway of rumination and seemingly uncontrollable agitation.

Serotonin regulates communication between various parts of the brain, and when it is imbalanced it can lead to impaired function in different areas of the brain that help regulate thoughts and emotions. Our emotional brain takes over rational thinking and it's the reason we may feel more inclined to act on our impulses and like we're out of control.

Sleep may be impaired when we're going through a difficult period, and even though I've suggested getting good sleep, I fully understand how hard that can be sometimes. One important note is that when we're experiencing loss and endings, we have a tendency to stay up later at night, actively impairing our sleep hygiene. We can try to take back control by implementing

HOW TO SUPPORT OURSELVES BY INCREASING SEROTONIN

EXERCISE

SUNLIGHT

TIME IN NATURE

GOOD SLEEP HYGIENE

FOODS HIGH IN TRYPTOPHAN

MINDFULNESS AND MEDITATION

GENTLE ACTIVITIES THAT MAKE US FEEL GOOD

healthy sleeping habits. You can read more about these tips in Phase 3.

Tryptophan is the building block that serotonin is made from and you get it from food. Increasing your intake of foods high in tryptophan (see page 248) can help replenish the building block needed for serotonin production.

Oxytocin

Oxytocin is often referred to as the 'love hormone' or the 'bonding hormone'. It is involved in various social and emotional processes, including promoting social pair bonding, trust and empathy. It's the hormone that makes you feel all warm and fuzzy when you cuddle your loved ones. Not just romantic loves – it's also responsible for social bonding between in-group members and showing a bias towards an out-group. What this means is that oxytocin is responsible for determining who is part of your close group and who doesn't belong in it.

Though oxytocin is mostly known as the love hormone, according to more recent research it's also known as the 'crisis hormone'. A study[44] showed that oxytocin increases during times of relationship instability and distress. Researchers discovered that lovers who were more invested in a relationship released more oxytocin when thinking about their lover compared to the person who was less invested. This explains that feeling of wanting to hold on so badly when we break up or lose someone. Oxytocin gives you that feeling of wanting to bond with the person you can't contact any more. If we're dealing with a break-up, checking in on them all the time and even having contact with them can slow down this process.

Oxytocin is the bonding chemical, and the only way to heal that is time. The brain needs time to undo the bonds it previously made.

If you're dealing with the death of a loved one, then of course you do not want to undo any bonds, but you need time to be able to allow someone else into your life. You need time for the meaning of those bonds to change. If you've lost someone in this way, it's hard to imagine that the painful feeling will ever go away, but over time you will start to build new experiences on to your memory of that person. You will never forget them, and you shouldn't try to like you would with a break-up, but there will come a time when you'll begin to feel better. Your life will grow and continue to shape around the memory of the person you lost. Your brain will make new pathways and associations that will honour their life as you begin to experience joy and purpose on your own. As time goes on, you'll start creating a new narrative around who you are, and your meaning in life will help make sense of the loss and the grief. This narrative will determine where you're going in life and what this person meant to you, and you'll honour the past while moving forward into the future. Losing someone means you lose a little part of who you are, but over time you'll build a new story and identity.

If you've lost a dream, it will take time to recalibrate towards a new goal. Chapter 11 discusses failures and growth mindset. But for now, I want you to bear in mind that even though your brain is suffering right now, it's building resilience for the future. Don't let the knock in your confidence bring you down. Your brain is a remarkable and adaptable piece of machinery that can handle a lot more than you give it credit for.

Regardless of the loss you're experiencing, to aid the healing process we must create a sense of safety and comfort. Doing so, especially with the support of other people – friends, family, support groups or therapy – can help us release oxytocin, which will give us a sense of connection and support to help us alleviate the feelings of loneliness and distress. The way I see it, it's like creating a soft cushion of support around you that you can lean on. Your brain will start to rewire itself with new meanings and experiences as you go through the passage of time dealing with this feeling.

As humans we need to feel wanted. Leaning on your friends and loved ones can help, but don't get into the trap of falling into meaningless relationships for the sake of feeling wanted by someone. This may give you temporary relief, but eventually you'll feel empty again. My suggestion is to break off all contact with your ex, with your previous boss and with any other relationship that doesn't serve you. Engage in valuable relationships. Perhaps those relationships are romantic, perhaps not, but ensure they make you feel good about yourself and are helping you move forward instead of pulling you backwards.

INCREASE OXYTOCIN
THROUGH MEANINGFUL CONNECTIONS AND MEANINGFUL AND DEEP CONVERSATIONS

Endorphins and endocannabinoids

Endorphins and endocannabinoids are natural painkillers that are created in the body. They are released in response to stress, exercise, pain and other activities such as laughing and crying. The word 'endorphin' means 'endogenous morphine' because it has morphine-like effects in the body. Apart from dampening pain, it also promotes feelings of well-being and euphoria, creating a sense of pleasure, reward and relief.

One of the biggest underlying themes with grief and loss is that our body goes into low-power mode. My suggestion is to use the self-care tools from 'Chapter 1: Break the Cycle' and incorporate the tools given here too. You'll see there's a big overlap and that the tools blend in with one another. For example, I discussed hobbies and have suggested taking up hobbies and engaging in exercise in this chapter. I know that exercise can be daunting for a lot of people. Chapter 12 discusses exercise in greater detail, which usually helps people understand why it's so beneficial. But if you don't feel up to exercising, my suggestion is to engage in social activities that make you smile. Perhaps you can watch a feel-good film or a documentary about the planet. Sir David Attenborough always knows how to make one smile.

WAYS TO RELEASE ENDORPHINS & ENDOCANNABINOIDS:

EXERCISE

RUNNING

STRETCHING

DANCING AND SINGING

LAUGHING AND SOCIAL CONNECTION

HEAT EXPOSURE (SAUNAS AND BATHS)

HOW TO DEAL WITH ENDINGS AND LOSS

Emotional expression is one of the best ways to deal with endings and loss. Studies[45] have shown that expressing ourselves through art and creativity can help us develop coping mechanisms to manage grief.

I was twenty-four when I went back to ballet and started from the bottom. I had done a bit of ballet in school when I was a child but dropped out when I started being bullied. At twenty-six, I was even more drawn to it when I moved to my new city and lost my relationship, my life back in London and my dream of becoming a medical doctor. Dancing gave me an outlet to be able to express all the things I had lost while building something new. You know that movie scene where you're the main character of a sad drama but they end up being OK in the end? That's how I felt. Art gives us an opportunity to take a break from the ruminating dark thoughts and intense feelings. Art enables us to see things from a different perspective.

George Michael wrote his album *Older* after losing both his lover, Anselmo Feleppa, and his mother to cancer. *Older* was one of the most powerful statements that Michael made in his career because through his lyrics he was coming out to the world about his sexuality. When Feleppa died in 1993, Michael felt as though he would never find creativity and inspiration again because he was so deep in grief. Less than a year after Feleppa's death, Michael sat down at a piano in his Notting Hill flat and was shocked to find himself writing an entire ballad, 'Jesus to a Child', in just two hours. The lyrics addressed the secrecy of his and Feleppa's relationship. He performed the song barely a week later at the MTV Europe Music Awards. The interesting thing about the album is that amid the melancholy and sadness there are a few songs on there that are more joyful and happy, like 'Fastlove' and 'Spinning the Wheel', which shows us that it's OK to be playful and explore another perspective of grief.

I believe that there is an artist in all of us, and while not all of us can be as talented as George Michael, we can certainly try to create something from our grief. It may be drawing or painting, or perhaps you may want to try to write a song. I leant on ballet dancing during my grief.

WAYS TO EXPRESS YOURSELF
DANCING | SINGING | POETRY | ART |
SCRAPBOOKING | JOURNALLING | HOBBIES |
DRAWING | PLAYING AN INSTRUMENT |
LEARNING AN INSTRUMENT

Over time the brain rewires itself around the grief to accept a life without what was lost

Sense of comfort

As humans we need to feel safe. Safety is one of our most basic needs. When our basic needs are not being met, it can have a significant and detrimental impact on our physical, emotional and psychological well-being. Basic needs are the fundamental requirements that are essential for our survival and overall health. When our safety needs are not being met because we do not feel safe in our environment, physically or emotionally, we can experience high levels of stress, anxiety and fear, which can result in chronic stress-related health issues.

Endings and loss can really upset our basic requirement for safety. Therefore my suggestion is to tighten up on the things that bring you a sense of safety and security.

- Community support
- Staying at home in your comfort zone
- Living with friends and family
- Calling friends and family
- Financial safety
- Bereavement support
- Mental health support
- Journalling
- Maintaining routines
- Getting professional help
- Self-care
- Surrounding yourself with people who are understanding and empathetic

- Limiting contact with individuals who are insensitive or unsupportive

Remember that healing from loss is a process that takes time, and there's no right or wrong way to feel. It's OK to seek help as you navigate your feelings and work towards feeling safe and secure again.

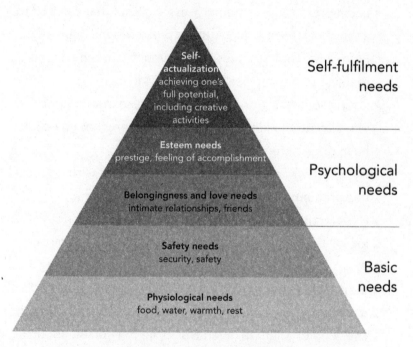

Maslow's Hierarchy of Needs

Your brain will rewire itself to adapt to a new reality and you will slowly start to find acceptance of what you've lost.

REWIRING ESSENTIALS – ENDINGS, LOSS AND GRIEF

- Studies have shown that the brain reacts to the pain of a loss in the same way that it would to physical pain. There is even some research to show similarities to the pain of breaking a bone.
- Loss can mean that we are grieving people, and sometimes it's grieving lost opportunities.
- Endings can cause extreme stress and grief that can lead to a range of issues, including sleep problems and depression.
- From a neuroplastic perspective your brain is having to readjust to a completely new way of living.
- Losing something or someone can cause insomnia, so you stay up thinking and ruminating over what you could have done differently.
- Hyperactivity of the amygdala is associated with impaired emotional processing and depression.
- Hyperactivity in the emotional centres can override logical thinking, which can make us act on our impulses.
- Areas of the brain involved in motivation and craving are more active during a break-up. This can make you want to act upon your impulses and perhaps message your ex, check up on them or, even worse, lie to yourself about why you should get back together with them.
- Loss and endings hurt because we have lost something, but within it we have also lost our routines, habits, automatic reactions and behaviours.
- Tools to alleviate symptoms of loss and grieving – page 150.

NeuroToolkit 1: How to Ditch the Negative

The laws of neuroplasticity state that when one neuron drives the activity of another neuron within a specific time frame, the connection between those neurons becomes associated and is strengthened. What this means is that you have probably learned to associate particular thoughts so that triggers and events spiral you into a ruminating abyss by neuronal association. The neural basis of morality is still a topic of ongoing investigation. However, we do know that, regardless of whether certain behaviours are socially and morally right or wrong, your brain will divert to the pathways that have been strengthened through repetition. That said, a trigger doesn't necessarily need to be something negative. For example, a common one is you're more inclined to go to the gym from the office than if you went home first. Or perhaps you have friends who influence you positively and make you feel really good about yourself. A trigger is essentially a cue to remind us to do something. These cues can be environmental, social, sensory, time-dependent, driven by a preceding event, etc. For the purpose of this exercise we're going to identify negative triggers. But before we do that, I want to entertain an idea about repetition and plastic change.

**When we break down neuroplasticity
to its most basic form:
repetition + attention + intention = lasting change**

I named my dog after Kobe Bryant, one of the greatest basketball players of all time. Just like Bryant, my dog Kobe loves to play ball, hence the name. Kobe Bryant is a huge inspiration to me and is a source of inspiration to many others from all walks of life. Bryant was known for his incredible work ethic and dedication to improving his basketball skills. He was rumoured to have taken hundreds or even thousands of shots per day during his career, especially during the off-season and when he was younger. However, the exact number of shots he took per day varies depending on the source and the specific time in his career. In one story Bryant's trainer saw him sweaty and alone on the court before team practice. Stunned by this, he asked him when he finished his training, and Bryant replied, 'Oh, just now. I wanted eight hundred "makes". So yeah, just now.' Kobe wanted to take 800 shots before the team practice even started.

In another interview Bryant said he used to make 1,000 shots a day when he was a teenager. Perhaps as he got older and more experienced, he adjusted his training routine. Assuming that is the case, that he started out with 1,000 shots *per day* and then went down to 800 per day, becoming one of the most successful players of all time in the process, tells us about the power of repetition. Bryant deliberately practised shooting at the hoop until his brain was wired in a way that it became automatic. Mental heuristics will have made subconscious decisions for him when deciding when and how to shoot a ball into the hoop, shortcutting the decision-making process. It's a process that was deeply ingrained in his neuronal pattern: from having to gauge and decide how far away the hoop was, all the way through to his motor response ensuring that he threw the ball with the correct trajectory, at the right speed and right angle, so that it could land in the hoop. We never take into consideration all these processes when looking at a phenomenal player like Kobe Bryant. All these neuronal communications happened at light speed to ensure that he took enough steps with his lower limbs while his upper body shot the ball with the right amount of force and precision for another make.

I have only ever messed around playing basketball, but whenever the ball lands in my hands I go into sheer panic when having to take a shot, which I obviously miss. It feels like I have to make a hundred decisions before I can even attempt at shooting the ball correctly. I have recently taken up tennis lessons to fix the lifelong mistakes I've been making as a self-taught amateur, and it's similar. There are so many things I need to think about in order for me to hit the ball well over the net so that my coach doesn't yet again yell, 'Come on, Nicole!' in a disappointed tone. I know that at some stage the decisions I make as the ball approaches me will become decision-less through heuristics. My brain and body will make decisions with automaticity based on previously ingrained experiences. But until then tennis is more of a mental workout for me than physical. Which is saying a lot because after every lesson, and the paces my coach puts me through, I walk back home uphill, waiting to collapse when I get through the door.

Bryant didn't have to think of any of that, because the automaticity of his brain unconsciously already knew exactly how to respond to any opportunity at a shot. Bryant made those shots with deliberate intention. He woke up every day for practice (attention) and repeated the same grind day in and day out. Kobe's formula for success was repetition (800 shots per day) + attention (showing up every day) + intention (intentionally shooting the ball every day for success) = an incredible player.

Now imagine how many 'makes' per day you've had with negative thoughts. The results from a 2020 study[46] that was published in a highly respected journal concluded that on average we have around 6.5 thoughts per minute. That's approximately 6,000 per day if you sleep eight hours a night. Of course, this number may vary and it is only one paper, but it's opened up the door for additional research. Either way, we have a lot of thoughts every day. So if Kobe Bryant took at least 800 shots a day and became one of the world's greatest basketball players of all time, then think about how good you have become at speaking negatively to yourself and focusing on all the negative things in your life if that's what you've been repeating. I mention this not because I want you to feel like this journey of rewiring is daunting but because I hope that it inspires you to stop talking to yourself so badly. I hope that it inspires you to stop looking at all the negative things in your life so that you can stop repeating this pattern and solidifying it even more. If you've practised responding to triggers negatively, I'll say it again: your brain doesn't know right or wrong; it just knows that particular pathways have been strengthened through repetition. The triggers and responses may be negative for now, but just as we can create a strong association for negative responses, we can also create strong associations between positive responses so that we can live more like Kobe and strengthen our road to success.

Identifying our triggers is important because our brain likes to stay in control within its comfort zone. When these triggers arise and we have prepared for them, we are better equipped to deal with them instead of succumbing to the type of behaviours that we then beat ourselves up about.

> '*Between stimulus and response lies a space. In that space lie our freedom and power to choose a response. In our response lie our growth and our happiness*' – Unknown

Usually a cue is something in your environment or something you see or feel, which then triggers a habitual behaviour that is reinforced by a reward. That's your general habit loop. But these cues can also trigger us to feel a particular way and send us down a spiral of negative thinking. So for the purpose of this next exercise, we'll identify cues that lead us to negative habits or negative thinking and rumination.

HOW TO USE THE LAWS OF NEUROPLASTICITY TO STOP RUMINATING THOUGHTS

LET'S BREAK THESE DOWN.
1. SIGNALS FROM YOUR ENVIRONMENT (CUES)

Your surroundings drive your habits, behaviours and actions. For example, you don't normally crave soda, but the minute you step into a restaurant, the first thing you think about is wanting to order a Diet Coke. Or perhaps you drink more coffee when you're at work because it's free there. The environment you're in cues your brain to want those things that you normally wouldn't when you're at home.

CAN YOU IDENTIFY THE ENVIRONMENTAL CUES THAT TRIGGER YOU TO RUMINATE?

2. SOCIAL SITUATIONS AND SETTINGS

Social cues can be the company you keep or the social setting you're in. Perhaps you have a group of friends who influence you positively and another who you want to go to the bar with. Maybe you have a toxic friend who triggers you to gossip or you feel rubbish about yourself when you're around them.

CAN YOU IDENTIFY THE SOCIAL CUES THAT TRIGGER YOU TO RUMINATE?

3. SENSATIONS AND SENSORY TRIGGERS

You may be triggered by something you see. For example, after seeing photos on Instagram of other people's bodies and journeys, you might make a negative comparison which sends you into rumination. In a world of rampant social media use and constant images of how much better other people are doing, I would imagine that this cue is one of the strongest driving factors to rumination.

CAN YOU IDENTIFY THE SENSORY CUES THAT TRIGGER YOU TO RUMINATE?

4. WRONG TIME, RIGHT PLACE (TIME-DEPENDENT CUES)

This cue is also quite common. A lot of people I know and talk to say that they tend to ruminate the most when they're about to go to sleep. This also applies to people who are trying to stop habits such as smoking and drinking, and as soon as the clock strikes 6 p.m., the urge to do either becomes unbearable. Your brain may have made associations with some or more of these cues which have led you to fall into a habit of ruminating. A habit that now happens on cue. The good news is that by having a plastic brain, these associations and ruminating thoughts can be undone by dismantling firing patterns.

DISMANTLING THE FIRING PATTERN

Now that we have identified potential triggers for your ruminating thoughts, we can work on dismantling these firing patterns. A lot of the time these thoughts happen with automaticity, meaning that they're performed unconsciously or involuntarily as a reflex, innate process or ingrained habit. This means that your neurons are firing in a sequence that's been practised so often that it fires without you even realizing it, perpetuating your beliefs and your thoughts, which then dictate how you approach life.

Your neurotoolkit

Step 1: acknowledgement and understanding the brain

When you find yourself ruminating, the first step is to acknowledge that it's happening and that there is nothing wrong with you and your brain. It's not you, it's your brain. It's just doing what it knows best.

On a neurological level we need two things to make changes as adults: *attention* and *intention*. In the introduction we discussed how the child's brain can make changes easily, but we need to tell the adult brain what's important. The brain areas of the RAS and the SN are responsible for ensuring that we shine a spotlight on what's important to us. Attention means that we're telling the brain *this* behaviour is unwanted and we need to make a change to it. *Intention* is putting in the work to make that change so that we can *rewire* our brains to stop repeating the negative self-talk. We can select where we put our attention and we can select which information we want to pay attention to. Our brains are capable of doing this while simultaneously suppressing irrelevant or competing distractions.

Acknowledging that we have this power can give us a sense of relief because we often believe that we are a product of whatever thought arises and that we need to run with it. But we have the power to change this. Acknowledge this and acknowledge that your brain is only doing what it's done best for so long.

It's time to change, though. You can and you will.

Step 2: leverage cues and triggers to make a change

Knowing that the brain needs cues and triggers to be reminded to do something means we can leverage this mechanism to our advantage.

Think of a positive habit or behaviour that you're trying to implement. My suggestion is the physiological sigh (see page 41). If you begin to intercept negative thoughts with the physiological sigh, this will weaken the firing pattern of associated neurons. When you disrupt the firing pattern, you can start dismantling it. By interrupting the thoughts with a new positive habit, we can use the negative thinking to remind us of the new habit.

This allows you to turn the negative into a positive.

Being in a state of high emotional arousal when ruminating can make it hard to think clearly. Our emotional brain takes over and our neocortex, the part responsible for making analytical and factual judgement calls, takes a back seat. This means that intense emotions drive feelings, and narratives may even be exaggerated. The physiological sigh can help you regulate your nervous system back to a calm state so that you can analyse your thoughts with a more level head.

Step 3: analysis of thoughts: reflecting and reframing

Emotions have a significant effect on the way we think and make decisions. Have you ever refused to go to a party because you hated all your clothes and thought everything looked awful on you? I have a few recollections of having some outbursts regarding this very topic. That's our emotional brain making decisions. Usually you look back on photos in hindsight to realize that your outfit wasn't as horrid as you made it out to be. Your negative self-talk and beliefs might be leading you to misinterpret information about the world around you. It's important to respond and reflect on your thoughts with a logical mindset, one that is driven by factual and non-emotional evidence.

Reflecting on your thoughts when you're feeling less emotional is an important step because it means that you start to perpetuate a more factual truth about yourself because you're addressing the situation with more rationale. The frontal cortex is more engaged and the limbic system isn't overriding your decision-making. And when negative thoughts arise in the future, you'll be better equipped to deal with them and you'll be more likely to revert to this state of logic more quickly. By assessing our thoughts more rationally, we start to anchor them to something that is self-directed through metacognition, rather than being stuck in a loop that's unstructured and circles around in your head.

Journalling

Another great tool for gaining clarity on our thoughts and emotions, which is well grounded in science, is journalling. Journalling helps reduce the emotional load and diminish the frequency of ruminating thoughts over time. This is because when we journal we have to use complete sentences, unlike when we are ruminating.

Journalling also helps us create a narrative around the issue that's upsetting us, and this is a perfect example of when creating a narrative can be useful. Writing something out with a coherent narrative can help us focus on the positive aspects of our lives while helping reduce anxiety, rumination and the emotional load we carry when engaging in negative self-talk. Journalling also activates the areas of the brain involved with generating words, which steers our thinking pattern away from the DMN as it requires more energy and recruits larger brain networks.

Through the act of writing, we start to externalize the

ruminating thoughts and make them more manageable, which in turn makes it easier for us to address them. Journalling allows us to confront our thoughts and feelings in a controlled environment, which can be less overwhelming than the constant background noise of anxiety or rumination.

Step 4: don't beat yourself up about it

When I spoke to a psychiatrist at Brown University who works in addiction, he told me that the most important step to changing any behaviour is to not beat ourselves up about it when we do the thing we're trying to stop. In this case when you start ruminating. Firstly, it is in no way helpful, because if it worked, we wouldn't have any negative habits at all because simply beating ourselves up would have worked the first time. Secondly, it can make things worse because when we beat ourselves up, this gives us a sense of control over the situation, which further reinforces the behaviour. Subconsciously you know you don't need to work on addressing it, because you can just do it anyway and beat yourself up about it later. Creating a sense of control over the situation also activates reward centres in our brains, which make us feel good temporarily but don't fix the root issue. Lastly, you're not changing anything because you're ruminating over the fact that you were ruminating, so you're just going around in circles anyway, attempting to gain control over it while actually getting nowhere.

What should you do instead?

When I spoke to Dr Jud Brewer, the psychiatrist at Brown University in addiction, he suggested analysing your thoughts instead. Understanding why you were triggered to do that negative thing in the first place can help you navigate those triggers more effectively.

We have discussed the brain's negativity bias and how we tend to avoid negative information and don't learn from negative reinforcement such as berating yourself. We've also learned that repeating negative thoughts induces a freeze response that keeps us in a repetitive loop instead of fixing the issue. Therefore beating yourself up is not going to work.

Science shows that using positive reinforcement is the best option when trying to make a change. Congratulating yourself for the small wins and acknowledging them. Since we tend to dwell on the negative, we sometimes fail to see how far we've come. But making sure that we celebrate the small wins, and staying away from berating ourselves, appears to light up reward centres that can help solidify a new behaviour.

QUESTIONS YOU CAN ASK YOURSELF

Are these things you're saying to yourself really true?

If they are true, can you change them?

If you can't change them, can you accept them?

If you're feeling anxious, what is your body trying
to tell you?

If you're feeling sad, can you find compassion for
yourself?

Does this thought help you take effective action towards
making a change?

What new story or thought can you focus on now?

How can you see this in a different way?

What can you take away from these thoughts?

What steps can you put in place to ensure you don't feel
like this again?

Just because you've compared yourself to someone on
social media, does it mean you're any less?

Can you stop comparing and meet yourself with compassion and gratitude for the things you already have?

Phase 2

Shift Your Narrative

Looking back is good for learning. It teaches us but dwelling on it does not serve us. Look forward in the direction of life.

Rewire Your Subconscious

1. Leave your phone alone
2. Visualization & attention
3. Repetition
4. Make space
5. Push through the boundaries
6. Create a strategy & prepare for setbacks
7. Step through fear & conquer self-sabotage

life (noun)

The condition that distinguishes animals and plants from inorganic matter, including the capacity for growth, reproduction, functional activity and continual change preceding death.

You see, by definition we are designed to evolve . . . *continual change preceding death.* So when did you begin to believe this wasn't a possibility? Because it's a fact of life that we have the capacity for growth.

Life is a story that we get to write. Create it. Don't settle for it.

Phase 2 has all the tools you need to create the life you want.

HOW TO REWIRE YOUR SUBCONSCIOUS:

IT IS ESTIMATED THAT THE MAJORITY OF YOUR BRAIN (AROUND 90%) IS SUBCONSCIOUS AND IT'S DRIVING YOUR EVERYDAY ACTIONS AND DECISIONS.

Rewire Your Subconscious

Most of our brain operates on a subconscious level, meaning we are on autopilot without conscious thought. You're not telling your heart to beat every second or consciously telling yourself to blink and swallow, nor are you thinking about the way you cross your legs when you're deep in thought, working at your computer or studying for an exam. You don't pay attention to your mannerisms either. For example, when you first learn to drive a car, you're very aware of your actions, constantly checking your mirrors and your blind spots, but eventually, once you're skilled enough and you've repeated these actions enough times, thinking about which pedals to press becomes an automatic part of how you behave and so you do it almost subconsciously. This is a positive thing because if you had to pay attention to every single thing you did all the time, you would be exhausted by lunchtime. Nearly all your thoughts, actions and decisions are conducted in different regions of your brain on a subconscious level. Actions and thoughts that you have repeated many times are now automatic. Your brain is constantly making decisions for you on a subconscious level, cross-referencing previous experiences to decide the best course of action. We've mentioned this before when talking about experience-dependent plasticity. This is the brain's ability to fine-tune its connections and functions based on how you interact with your environment. Remember in the introduction when we spoke about how our brain creates a database of information based on your experiences? Think of your subconscious brain as a web of probabilities and out-comes, and when you need a solution to a problem your brain quickly checks the possible algorithms for a result.

The same thing has happened with your habits and behav-iours. They have been preprogrammed, and if you are somebody who repeats a negative narrative to yourself, then based on your inner beliefs this will drive many of your daily actions. If you are somebody who's been put in a box by your family and peers, or someone whose confidence has been knocked due to an abusive relationship, my promise to you is that you can break free. We can change our narratives and we can change those subconscious behaviours so that we can become someone who walks into a room confident and assured instead of shy and worried.

But I want to put forward an idea. If your brain has built an entire database of information based on your experiences, then surely you should give your brain a bit more credit when you experience something for a third, fourth or fifth time? For example, when you first start online dating you feel like a rookie who doesn't really know how to handle the rejection of not being asked out on a second date. By the second, third or fourth time you should know that this is how it goes, and in theory you should be equipped to deal with it. We tend to focus on the negative and think there is something wrong with us, but what if we saw it as yet another piece of information on how to handle a situation? A lesson from which we can learn something so that we can be more resilient the next time. Can you apply this to other areas of your life?

1. Leave Your Phone Alone

Spring-clean

The information we consume on a daily basis has a big influence on us. Many of us get this information predominantly through social media. But it can also come from other forms of media, including the TV shows you watch, the magazines you read and the news you engage with. Unfortunately a lot of these channels often promote harmful messages, conveying unrealistic expectations of what someone should look like or the achievements one should accomplish. Though the content we consume has a major influence on us, the good news is that to some extent we can control what we consume and focus on content that's inspirational instead of negative for our self-image.

In a survey conducted by Common Sense Media in 2018, 70% of teenagers reported feeling left out when they saw others posting activities online that they had not been invited to. Studies[47] show that young adults who spend more than two hours a day on social media are twice as likely to feel socially isolated compared to those who spend less than thirty minutes a day on social media. Another study[48] found that adolescents who spend three or more hours per day on social media are more likely to report high levels of depressive symptoms compared to those who use social media for less than one hour per day. The American Psychological Association (APA) conducted a survey in the same year that found that 45% of respondents reported feeling anxious due to the perception of others' lives on social media. And according to the National Sleep Foundation in 2021, 67% of Americans reported using electronic devices like social media before bed, which is associated with poor sleep quality. Multiple studies have strongly linked heavy social media use with an increased risk of depression, anxiety, loneliness, self-harm and even suicidal thoughts.

If you follow accounts on social media that make you feel bad about yourself, or which make you think about what you don't have and contribute to negative self-talk, then my suggestion is to start muting or unfollowing those accounts. Anything that leads you to comparison or gives you a negative feeling about yourself should be assessed. We should be in control of what is affecting our mental state.

This isn't to say that all content is going to affect us negatively. For example, you don't need to unfollow people who portray unrealistic bodies and unattainable lifestyle choices if they don't affect you. My only suggestion is that you pay careful attention to the content you're consuming because it has a large yet subliminal impact on how we perceive the world either consciously or subconsciously.

Unfollow	Follow
Accounts that make you compare yourself	Accounts that make you feel good about yourself
Accounts that feed in to your negative self-belief	Accounts that educate you on your negative self-talk
Accounts that trigger negative thoughts	Accounts that help you identify your triggers
Accounts that reinforce negative beliefs about yourself	Accounts that reinforce positive beliefs about yourself

Do not use your phone first thing in the morning

During the transition from sleep to wakefulness your brain is in a relaxed mental state. Your brain gradually shifts from the slower-wave brain waves to higher brain frequencies as you become more aware of your surroundings but are still in a somewhat drowsy or relaxed state. You'll find that you're in a more dreamy, meditative and creative state of mind, opening up a world of possibility.[49] But by grabbing your phone first thing in the morning, you're skipping this essential and wonderful brain state. You have all day to be on your phone – why not capitalize on this special moment of peace and tranquillity in the morning?

Checking your phone and exposing yourself to potentially overwhelming information and notifications can disrupt this relaxed state and shift your brain into a more alert and stimulated mode. Taking time in the morning without your phone allows for a moment of mindfulness, introspection or reflection on the day ahead, aligning with the contemplative aspect of being in a dreamy wakeful state, whereas doom-scrolling first thing promotes reward-seeking[50] behaviour early on. This means you'll be chasing quick rewards for the rest of the day, making it harder to practise good phone hygiene when needed, like when you're at work or in conversation with your friends over coffee. Engaging in a morning routine that doesn't involve immediate phone usage encourages a more relaxed and focused start to the day. Additionally, excessive use of social media has the potential to overload your brain. Constant streams of updates and notifications can have a negative psychological impact on your brain. If we think about brain energy as mental currency, you're essentially giving it away early on and risking cognitive overload for the rest of your day. Why not hold off on these effects until later?

It's important to be mindful of the information and stimuli we expose ourselves to in the morning, as it can shape our mindset and outlook for the day. This is the perfect time of the day for a routine that takes advantage of this suggestible state of mind to engage in self-care, meditation or other mindfulness practices such as visualization (which is discussed in the next step) to instead set a positive tone and intention for the day ahead.

2. Visualization & Attention

American swimmer and twenty-eight-time Olympic medallist Michael Phelps began visualizing at the age of eleven. His teachers were exasperated with his inability to focus in class and had little hope for his future. But Phelps was fortunate enough to have Bob Bowman as his swimming coach, who gave him and his mother a book on relaxation techniques.

Phelps quickly learned to visualize, which taught him to be calmer overall, but more importantly he learned to visualize himself being a better and faster swimmer. He would visualize how he wanted race day to go, with each race going well and him staying confident under pressure. But more importantly Phelps would also visualize things going badly, because not everything is under our control all the time. He would run through these scenarios and ask himself questions like, 'What if things don't go well?' This allowed him to enter race day with a calm mindset because he had visualized all possibilities and outcomes, so that he wouldn't be thrown off his game if something unexpected happened. Remember, the brain likes to be in control, and anxiety arises from not having control over certain situations. By mentally rehearsing such outcomes, he would be prepared for whatever the day threw at him. His coach even said, 'By the time Michael gets up on the blocks to swim in the World Championships or Olympics, he's swum that race hundreds of times in his mind.'

The visualization skills that Phelps used during his career made him one of the best swimmers of all time. At fifteen he became the youngest male swimmer since Ralph Flanagan in 1932 to qualify for the US Olympic swim team. He's broken world records and holds the all-time record of winning eight gold medals in one single Olympic Games. When it comes to mental training and visualization, his coach says that he's the best he's ever seen.

Bear with me for another Djokovic story; I think you've probably realized by now he's one of my favourite sportspeople. He talks about practising visualization too. Djokovic harnesses his mental practice to visualize the outcome of a match. What I love most about Djokovic is his metamorphosis throughout his career. Many people are of the opinion that he's arrogant, obnoxious and ill-tempered on the court, or at least that he used to be, and they can't see past that version of him. But Djokovic has openly discussed his anger on court and adopted practices such as mindfulness, visualization and spirituality to harness his inner calmness and really leverage his mental strength for the benefit of his game. Professional coaches believe that all top athletes are physically capable in their given sport. They all practise the same shots. They all train the same amount and they all work hard. But what makes the best athletes better than the others is that they have the mental strength to push their boundaries. They strengthen their minds – not just their physical bodies – to improve their game and be better than their competitors.

When we bring our brain state down to a calm and relaxed state, especially if we can go 'deep' into meditation/self-hypnosis or visualization, our brain can reach alpha and theta brain waves. Alpha and theta brain waves are linked to a state where the brain is receptive to learning and forming new connections. It's a state where the chitter-chatter disappears and you get clarity about what you want to change. During such states the brain may be more open to creating and strengthening synaptic connections, potentially aiding in the process of learning and adapting.*

The Science of Visualization

In Phase 1 you learned about the importance of attention when making a change. You also learned about the power of our thoughts and how they can evoke plastic changes. The piano experiment (page 107) showed how new pathways can be created through thought alone. The milkshake experiment and the

---

* It's essential to note, however, that neuroplasticity can occur at various brain-wave frequencies and is not limited to alpha and theta states. Neuroplasticity is a complex process influenced by various factors, including brain-wave states, the nature of the activity or experience, repetition and individual differences.

reframing of stress study (pages 106 and 109) showed how our perceptions and belief about something can influence our physiology. These experiments showed us how important our beliefs are, and how these beliefs can elicit a physiological response and make us react differently, which is down to our hormonal response. If our thoughts are this powerful, and we know that they can induce a plastic change, then I hope you're starting to see why visualization could be an effective method for change.

Visualization is a technique that is gaining more popularity now, but it has been used in sports for learning since the 1984 Olympics. Many athletes, Phelps and Djokovic among them, visualize being successful in competition and use mental imagery to condition the brain for successful outcomes. Research[51] shows that they stimulate the same regions of the brain as when they physically perform the same movements and actions. Mentally rehearsing your performance, and through repetition, means that the actions become habituated in your mind.[52] If you've already imagined executing something, then it will be easier to execute it later. I like to call visualization mental training. Much like with physical training, in which we repeat movement to get our muscles and cardiovascular system stronger, with mental training we're strengthening our brains to adopt particular behaviours and improve learning-based outcomes.

Visualization needs to be combined with real-world experiences, meaning that you can't mentally rehearse something and not execute it in real life because visualization cannot replace real experiences[53] . . . Then they would just be thoughts.

The mental rehearsing and the execution need to be in parallel in order for this to work. For example, you want to start running in the morning before work, but you always seem to hit snooze. One way to avoid this is by mentally practising your movements the night before. You mentally rehearse getting up in the morning and NOT snoozing your alarm; you also rehearse knowing that you'll be tired and want to snooze. We need to take into consideration that these things will happen instead of pretending that they won't. You can then visualize yourself getting up, tired, groggy and wanting to go back to sleep, but you keep putting one foot in front of the other, getting out of bed and dressed. When it comes to the morning, you'll need to execute it or you will have practised it in vain. Even if you never leave the house, practising the mental rehearsal until the end of the imagery (getting dressed) will mean that eventually that sequence of events will be stronger; then you can add on to the sequence, so you add that once you've got dressed you put on your headphones, choose your music, start your running app, walk to the front door and leave the house.

Pair visualization with attention

You need to bring daily attention to the thing you want to change, because otherwise your brain will revert to mental heuristics, falling back into a trap of rumination and negative self-talk because that's what your brain has practised and repeated for so long.

Adding in a mental training session to your day, ideally in

the morning, can help you execute your day accordingly. It can be used for self-limiting beliefs and low self-esteem by practising walking into a room/coffee shop/meeting room with your head held high; it can be used for practising staying calm before a presentation/interview; or it can be used to come to terms with loss of some kind. For example, if you're dealing with a break-up, you can mentally rehearse not checking their social media or not texting them, but doing something else when those urges arise. Mentally rehearsing that you're sad in this moment but you know that you'll be OK eventually will help too. We should never try to brush over what we're actually feeling, but rather understand and reframe our mindset to remind ourselves that we will be OK instead of catastrophizing thoughts and ideas such as 'I will never fall in love again', 'I will never find anyone who likes my flaws again' or 'I'm going to be alone forever'. These thoughts are not helpful and they lead us into a rabbit hole of low self-esteem.

What we should be practising instead is knowing we feel awful but being realistic about the situation . . . Will you honestly never meet anyone again? Of course it can be frightening to think that we may not, but the probability of that happening is low; it might, but we cannot repeat those things to ourselves, especially if we're hoping for a different outcome. We carry our beliefs into relationships, and a perfect example of this is when you've been hurt or cheated on. If you carry that belief into a new relationship, you'll enter it believing that relationships are doomed and that this person will deceive you too; you'll then start looking into things that aren't there, probably freaking this person out and pushing them away, perhaps even pushing them into lying to you to keep you happy. And now we have a self-fulfilling prophecy because they're lying to keep you happy, and you know they're lying, and boom, doom and gloom, the relationship is ruined. Our beliefs, mental rehearsal (visualization), memory templates, negativity bias and attentional bias (the reticular activating system and what we choose to see) are all working together to shape the way we interact with the world, and so the stories we repeat to ourselves are extremely powerful and we need to be mindful of what self-fulfilling prophecy we're shaping for ourselves by the things we say to ourselves.

The first step to making any change is acknowledgement. Acknowledgement brings the subconscious narrative that you want to change to your conscious thoughts so that you can work towards changing it. Sometimes we repeat these things so often that they become part of the background and are so deeply ingrained as a habit that we forget what we're even saying. We may not even believe it any more either, but it's still a habit that we perform every day, reinforcing our actions.

What about individuals with aphantasia?

Aphantasia is a condition in which individuals lack the ability to form mental images in their mind. Most people can close their eyes and see mental images, but individuals with aphantasia rely on other cognitive processes to represent and understand information. For those people, visualization techniques may not involve vivid mental images but they can still practise visualization in other ways.

- Instead of trying to visualize an image, focus on words and verbal descriptions. You could plan out your day using bullet points in a journal.
- Use mind maps or diagrams to organize and represent information. For example, you could colour-block your diary with reminders and visual representations.
- Create a narrative or story around the thing you're trying to visualize. You can do this by writing it down or speaking it out loud.

ACTIVITY

What are you acknowledging that needs change?

_____

_____

Write down the best time of the day for you to visualize.

_____

_____

What would you like to visualize?

_____

_____

Pay attention to the times you were happy. What were you doing right?

_____

_____

What were your habits? How were you approaching your day? Who were you around?

_____

_____

You can also find guided meditations and visualization practices online, and there are many apps that offer guided sessions.

3. Repetition

People always ask me how long it takes to make a change, and the answer is it can take anywhere from 18 to 254 days to create a new habit. This may sound demoralizing, but I want you to understand that this is not a temporary project. This is a new you, a long-term change that's going to undo all the years that have been backed by negative beliefs. You're dismantling the boxes your peers have put you in and the toxic traits you adopted through observing your family growing up, and undoing the microtraumas and the belittling that have lowered your self-esteem. This isn't a three-week transformation. This is the transformation you'll have for the rest of your life.

SMALL
HABITS
WILL BECOME
BIG HABITS AND
THEY WILL PAVE THE WAY
FOR A BETTER FUTURE, UNTIL THEY
EVENTUALLY BECOME WHO YOU ARE.

To strengthen a pathway and create a new thought or behaviour, the neurons that communicate with one another need to fire within a close time frame to ensure that they 'pair' and become stronger. They also need to be repeatedly communicating with one another to form a 'bond'.

They do this because when one neuron sends information to another, the receiving neuron gets better at receiving the information, and over time they communicate more effectively. Repetition is the most important aspect of creating new pathways for habits, behaviours and thought patterns.

The problem with making a change, though, is that usually we do it for a few days, but by the fifth, sixth or seventh day your brain is tired of holding this information in your conscious thought all the time and wants to revert to automatic so that it can be more energy-efficient. This is why it can be a good strategy to set daily reminders and intentions.

Remember, neurons that fire together, wire together.

HOW TO IMPLEMENT THIS CHANGE THROUGH REPETITION

DAILY MEDITATION FIRST THING IN THE MORNING

SURROUNDING YOURSELF WITH LIKE-MINDED PEOPLE

CREATING A MORNING OR EVENING ROUTINE

SUPPORT GROUPS WITH SHARED GOALS

CHECKING IN WITH A FRIEND

CHECKING IN WITH A COACH

SLOTTING IT INTO YOUR DIARY

SETTING INTENTIONS

SELF-HYPNOSIS

4. Make Space

The space between the trigger and the response will get bigger over time, and eventually, in that space, you'll make the decision to change your reaction. The space is the pause, the moment before you react. It's the opportunity to change your course of action and change the way that your brain pathways fire. At first this response is immediate, as if you have no control over it, but the more attention you pay to it, and the more you start observing it, the bigger the space will become. In this space you'll start to see that you have a choice in how you respond.

It's May 2023 and I have just moved to a new country. Everything is overwhelming. All the signs are in a new language. My car no longer feels like an extension of my body because I'm driving on the opposite side of the road. My memory templates aren't matching any of this with normality and my brain is in a state of constant alertness, taking it all in while feeling every decision and every action. I even have to think hard about shifting the gears of the car. My right hand is not used to doing this; it used to happen without thought. I am, quite literally, in overdrive. I can feel myself reacting negatively to almost everything that goes *slightly* wrong. But I decide to put my foot down. I can't let these overwhelming feelings take over and I most certainly don't want this to become a habit.

Recently I was standing in front of the receptionist of my gym. It was three minutes before my very hot, sweaty spin class was about to start, and she told me I could only use coins to buy water from the vending machine. *Surely this can't be right?* I could feel my emotions welling up, gearing up to spiral into negativity. But instead I pressed pause on the communication in my brain. I held back the two neurons* that wanted to fire in sequence to send me into a spiral, and I took a moment to make a decision. It felt like I was changing the gears on a train track and I chose to divert this potential trainwreck and get it back on track. I smiled and said, 'Any chance we can find a solution here? Perhaps you can lend me the coins from the till and you can put it on my account, and I'll bring the coins in tomorrow?'

The space between trigger and response was getting bigger and bigger. I was observing the crossroads of the routes I could go down: being annoyed and snappy with the receptionist OR understanding it wasn't her fault that that's the way the vending machine worked.

My biggest motto in life is SOLUTIONS NOT PROBLEMS. I believe that there is a solution to any problem, even if the solution is acceptance and detachment.

The space between trigger and response will increase over

* This is a dramatic effect. It's not two single neurons that drive those thoughts.

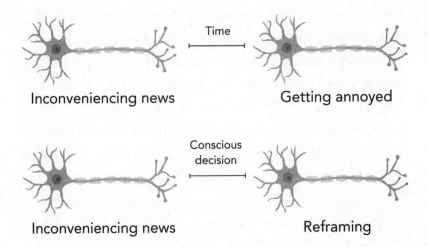

Inconveniencing news   Time   Getting annoyed

Inconveniencing news   Conscious decision   Reframing

time until you break that pattern and are able to make a conscious decision to reframe it instead. Don't worry if this doesn't happen right away; often we self-correct our behaviours without much conscious decision. The brain has a remarkable way of acknowledging errors and using them to encode new information for future planning. Again, this is thanks to experience-dependent plasticity and our ability to self-correct through acknowledging our errors in our brain. This means that slowly, as you start paying more and more attention to your actions, the space between that trigger and the response gets bigger. This won't happen overnight so don't beat yourself up if you do react instead of responding. As you work through these steps, the solution will become more apparent.

HOW TO MAKE SPACE
Practise patience in your life
Observe your reactions
Use the physiological sigh
Try a self-hypnosis routine
Make time for meditation

5. Push Through the Boundaries

How many times have you started a new habit and by week three you've reverted back to your old ways? Sometimes it's a conscious decision and other times you don't even realize until you've hit your head against the wall yet again. This is quite common, because the motivation and novelty of this new journey is fresh for the initial part, but thereafter this starts to dissipate and the brain reverts back to what it knows best. If you think of your brain as a big open field, your new habits and behaviours haven't been carved into any path yet. There is a nice concrete paved path with street lights and flower beds for the path that represents your old ways. These new habits have nothing – a dirt path at best; you have to create the path. You have to lay down the concrete blocks, plant the flowers and make it easier and more enticing to travel down. But when you get tired or you start forgetting about it, your brain will want to revert to taking the path it knows best.

Pushing through the boundaries is arguably one of the most important steps in achieving success in any habitual or behavioural change. There will inevitably come a time when you'll want to give up and revert to what is comfortable, even if what is comfortable is the wrong thing. It's one of the reasons people stay in toxic relationships or why patterns of abuse are repeated if someone comes from an abusive household. Pushing through the boundaries is the idea that you keep pressing on. You will get tired and you will lose motivation, but knowing this can empower us to keep making positive changes, knowing that things will get easier. This means you'll need to start relying on discipline instead of motivation to get you through this period.

Motivation can wane over time, but discipline is what keeps you going. People often place too much emphasis on motivation and usually give up when it disappears. Acknowledging that it will disappear at some stage gives us the power to focus more on discipline and consistency. Your brain is using a lot of energy trying to implement new changes, so you may need to adjust your schedule and make more mental space available by cutting out what isn't important. For a start, prioritizing the quality of your sleep is going to help. We'll examine this more in Phase 3, but what's important to understand is that sleep is your biggest optimization tool. All our new memories and learnings are consolidated during sleep. If you're not getting enough sleep, the chance of falling back into old patterns of rumination, irritability and negative self-talk is more likely. So it's a double whammy; you don't effectively store new memories and learnings, and you also might slip back into old patterns because you aren't functioning optimally.

Understanding that this is what happens can better equip you to deal with it. Remember not to beat yourself up about it if you do end up falling back to your old ways. We learned in Phase 1 that this is an attempt to gain control of the situation, but it doesn't work.

Another thing that I really want you to understand is that:

The brain is plastic, not elastic, and change isn't linear.

People often panic when they don't do something for a day, 'fall off the bandwagon' or feel they didn't do something right. But all the work you've put in is still there. Your brain isn't going to bounce back to its 'original' state like an elastic band if you're having a bad day. Perfection is not the goal. Sometimes we let our need to do things perfectly hinder our own progress or, worse, it stops us from even starting. As they say, 'perfect is the enemy of good'. Sometimes it's hard to see our progress and we're overly critical of ourselves, constantly wanting to be in place B and thinking about how we are not there yet. Slow down, take a breath and enjoy the moment. It doesn't need to be perfect. You're here, reading this book, doing the work. That's more than most people can say. I promise that all these little changes, new habits and new thoughts, all your efforts and all the work you're putting in, amount to something, even if you can't see it right now. The space between the trigger and the response is getting bigger and bigger, even if you have a bad day and react immediately. It doesn't mean you're back to your 'old' self. The brain doesn't work like that. Information input and pro-gramming is already in there somewhere; it doesn't just disappear. Trust yourself. Trust the process.

6. Create a Strategy & Prepare for Setbacks

Creating a strategy

Planning your journey so that you can stay on track can help on those days when you're feeling tired and unmotivated.

JOURNALLING

VISUALIZATION

SELF-HYPNOSIS

HABIT TRACKER

GUIDED MEDITATION

STICKING TO A ROUTINE

CHECKING IN WITH A COACH OR THERAPIST

CHECKING IN WITH A FRIEND OR SUPPORT GROUP

Some of these strategies have been discussed before – for example, visualization.

One of the best strategies for staying on track is to stick to a routine. We discussed cues on page 157 and we know that the brain needs environmental and time-dependent cues to trigger a response or a cascade of actions and habits. Additionally, when you cluster habits together, like in a morning routine, you are more likely to remember behaviours you've been trying to implement. For example, if you want to remember to take your supplements, you can use a habit-stacking method by putting them next to the kettle or your toothbrush or the fridge, which will remind you to take them. We can apply this strategy to behavioural change. Starting your day the same way and implementing a visualization practice, a journalling practice or whatever practice you want to adopt will remind you of what's important. Again, we need to bring attention to these things every day in order to bring about change. If you know you're going to have a hard day at work or you're tired, visualizing yourself being strong in those moments can help when you're faced with adversity. Knowing that you're in control will aid this journey.

These strategies can also include practising how to respond in a situation that you're used to reacting to in a particular way (page 168). Practising how you would approach a situation when it arises means that you're more likely to respond in a way that's been rehearsed. If you don't practise it, you may get caught off guard and you'll revert to your usual behaviour. But when you rehearse an alternate response, you have that card in your back pocket, ready to be pulled out in the heat of the moment.

Let's look at a more specific example. You're used to saying yes when people ask for your time instead of saying no because you're too busy. So, when someone asks you for something, in the moment you are caught off guard and respond with the answer you've practised all your life . . . yes. Your automaticity and heuristics dictate that that's what should come out of your mouth without even thinking about an alternate response. But if you practise responding in a different way, you're more likely to explore the idea of giving a different answer.

You could have these answers in your back pocket:

- I would really love to help out, but I am overwhelmed with work at the moment.
- My usual response is to say yes, but I am afraid that this time I just can't take this on.
- My schedule is pretty full until [insert time frame], but perhaps we could pick this up at a later date?

You get the gist.

Prepare for setbacks

Which brings me to my next point. Preparing for setbacks.

Sometimes on the journey of change and rewiring, we are faced with setbacks. These can come about as a result of our own internal state – being stressed, tired or moody on a particular day, which makes us more emotional and more likely to revert to old patterns. Often they're external factors that are out of our control, yet we attribute them to something being wrong with us. We begin to perpetuate the idea that we are failing because of external forces we have no control over. When we experience these setbacks without understanding why we're experiencing them, we start to believe that we will fail again further down the line. To add to this, our brain likes to be in control, so when setbacks arise we tend to panic. This can be due to our negativity bias, fear of failure or lack of autonomy, or it could trigger stress and anxiety.

Sometimes we even use these setbacks as an excuse because they give us a way out.

[On visualization and setbacks] 'It's how I wanted it to go, how I didn't want it to go and how it could go' – Michael Phelps

In 2008, Michael Phelps won a 200-metre fly race despite his goggles filling with water and having to swim blind. He recalled the experience in a later interview and explained that because he had visualized all the things that could possibly go wrong, he was re-laxed during the race. He swam blind for 175 metres of the race, won gold and broke the world record all at the same time. Phelps was prepared for the worst, including the fact that his goggles could come off. He knew that was a possibility and so he knew exactly how many strokes he had to take before tumble-turning and making his way back. He remained calm and in control of the situation. I think we can all learn something from him.

Studies[54] show that we can shift how we perceive these setbacks before they happen, to view them in a more positive light and as less of a detriment to our progress. This helps us respond more effectively when we encounter them, giving us greater determination and a renewed commitment to the goal instead of losing hope and giving up ... When we respond to setbacks with a more optimistic outlook, knowing we can overcome them, it ensures we stay away from ruminating thoughts such as 'I am weak' or 'I have no willpower' that contribute to our internal negative self-belief. Remember, the brain's negativity bias can exacerbate our beliefs about ourselves and amplify our negative narrative. But we can change our response to be more positive. Setbacks don't need to be the end of us. We can learn to overcome them and respond accordingly.

Many people ask me, 'What if I start ruminating when preparing for setbacks?' We know that sometimes our thoughts run automatically in our brains; they can be nonsensical and travel in our minds without any real direction. When we deliberately prepare for setbacks, it involves a practical and problem-solving mindset that helps us develop a plan that is outcome-oriented. This means that you're engaging networks in the brain responsible for executive functions that keep you in a state of logic and clarity. You're actively thinking about what might go wrong and how to mitigate those risks. In contrast, rumination tends to be more passive and repetitive, with a focus on negative thoughts and feelings without any clear solution.

SOME EXAMPLES OF SETBACKS:

LACK OF TIME MANAGEMENT

UNREALISTIC DEPICTIONS ON TV AND IN THE MEDIA

YOUR INTERNAL STATE: STRESS, TIREDNESS, HUNGER

SOCIAL MEDIA THAT MAKES YOU COMPARE YOURSELF TO OTHERS

CHILDREN AND FAMILY-RELATED RESPONSIBILITIES

FRIENDS AND COMPANY TURNING YOU INTO A DIFFERENT PERSON

FRIENDS AND COMPANY MAKING YOU FEEL BAD ABOUT YOURSELF

BEING SURROUNDED BY NEGATIVE PEOPLE

SUBSTANCE USE (SMOKING, DRINKING, DRUGS)

STRESSFUL EVENTS AT WORK

GETTING ANNOYED AT SMALL THINGS

So when we are preparing for setbacks, the primary goal is to come up with strategies and contingency plans to address potential problems. This puts us in a forward-thinking approach, which can help us feel more in control of the situation – unlike rumination, which lacks a clear focus on solutions and can perpetuate feelings of helplessness.

The game plan

EXAMPLE 1

You come home from work and you've lost motivation to cook and go to the gym.

Having prepared for this setback means you were anticipating feeling like this.

You're more likely to rely on routine and discipline and go to the gym or cook because you're more in control of your emotions.

EXAMPLE 2

Your children being late in the morning means they make you late for your workout class after drop-off.

This would usually result in you not going at all.

Having taken this into consideration and devised a plan B gym routine, you're more likely to go to the gym anyway and do a thirty-minute workout, even if it isn't as good as the gym class you wanted to go to.

EXAMPLE 3

You're going through a break-up and you're really hurt by the whole process. You didn't want to break up.

One day you see your ex on social media with a new person.

You'll probably still feel hurt, but you'll be better equipped to deal with the situation. This is because proactive preparation for this scenario can lower the impact it has on us.

Surprise is a strong emotion, especially when it's related to negative ones. Thus it requires a lot more emotional-regulation processes.

Preparing for this setback could look different to all of us. We discussed grief and how to boost neurochemicals relating to loss and endings on page 138, but some examples may be:

- acknowledging the pain
- the physiological sigh
- going for a walk
- calling a friend
- working out.

Setbacks arise due to internal causes such as stress and tiredness, or external causes which are out of our control. However, preparing for them gives us a sense of autonomy and a strategy to help us deal with them.

While preparing for setbacks involves recognizing potential negative emotions, it also emphasizes emotional regulation, encouraging proactive management and coping strategies. Unlike rumination, which amplifies negative feelings without direction, preparation is a grounded, realistic assessment that focuses on finding constructive solutions to risks and challenges and minimizing the anxiety associated with negative emotions. Preparing for setbacks minimizes the chances of us reverting to old patterns.

7. Step Through Fear & Conquer Self-Sabotage

The brain likes to be in control and feel safe, even if it is wrong. Self-sabotage is a fear-based mechanism that keeps us safe from the unknown. Our brains associate familiarity with safety, even if familiarity doesn't align with what we truly want. It's why people from abusive families can find themselves repeating the same patterns of abuse and why people repeatedly fall into romantic relationships with toxic people. That's what is familiar to the brain.

A known negative outcome is safer than the potential danger of the unknown.

This means you sometimes find yourself asking your friends for advice yet still do the complete opposite – even if you know what the right answer is and what you should do. Perhaps you make excuses for why you are a particular way because it's easier to do that than to admit you need to change. Procrastination and perfectionism are also forms of self-sabotage, and often mean you never embark on a journey of change because you're afraid you're going to fail (the root of procrastination) or you're not going to do it perfectly. These habits prevent you from making positive changes in your life and keep you in a loop of shame, because our brain defaults to staying where it's easier and safer rather than letting you step out of your comfort zone into the unknown.

But the reality is that change happens in the unknown, in discomfort. It's in the claustrophobic cracks between deciding to undertake something and actually going somewhere where things start to transform. You can't do this without feeling uncomfortable. Self-sabotage stems from fear, because we underestimate our capabilities; a known outcome, even if it's a negative one, is still safer than the potential danger of an unknown outcome. So we find ways to convince ourselves that we can't or shouldn't do something in order to justify the behaviours that keep us stuck.

Djokovic doesn't live without fear. He acknowledges that it's still within him. It's not that he's fearless; it's that he uses the fear as inspiration and he chooses to play anyway. It's in this fear that we create new synaptic pathways . . . where personal growth happens. The aim isn't to become fearless; the aim is to be brave. And when this happens, you'll come to realize that the world will commend your courage. Darkness still exists; you just learn to be the light in it.

'Henry Fonda was still throwing up before each stage performance, even when he was seventy-five. In other words, fear doesn't go away' – Steven Pressfield, *The War of Art*

Stepping through fear can go two ways

'Little by little does the trick'

You can dip your toe into the unknown and show your brain and body that you're safe. As you step into the unknown, you start to realize that things aren't so bad and so you'll be able to go a little further, telling your brain that you're capable of this change.

My other dog Max is a reactive little Malinois cross German shepherd rescue puppy who had a rather tricky start in life. Though she's only ten months old as I write this chapter, she's already bigger than your average dog and looks quite menacing. Max wasn't well socialized when she was younger, and she often tries to chase away people who get too close to us. To top it off, she also has a genetic trait that makes her nervy. She's very unsure of the world, scared of other people and

other dogs, so she reacts in response to that fear. You could understand owners of these kinds of dogs not wanting to take their dogs out in public, but then the dogs never have to face their fears and overcome their ingrained behaviours. Their owners might pull them away from people and other dogs, further reinforcing the idea that they are to be feared and avoided (remember the Ye'kuana tribe I discussed in the introduction passing on information from their body language?).

Because of my understanding of the brain and my obsession with neuroplasticity, I was determined not to go down this route with Max; I was excited to watch her change and adapt. I always take her out with me wherever I can; we sit and watch people go by and I reward her with treats in every scenario, especially when she makes the right decision to look at me or ignore people walking past.

We took her to the beach one day, where we go most weeks, and took her to her favourite pier that she likes to jump off. It's normally quite quiet, except that summer day there was a group of six people taking up much of the platform. Our initial response was to avoid the area and wait for them to leave, but we soon realized that they weren't going anywhere. So we mustered up the courage to go there, manage whatever situation might arise and persevere through the adversity. After all, it was a pretty good training opportunity.

When we got there we asked the group if they were comfortable with Max being around, which they were. Even better, they were compassionate towards her rather overactive behaviour and happy to be a part of her growth experience. By the end of the afternoon they were playing ball with her, and she was running between and around them without any problems. A new group approached and joined us as well, and Max was seemingly unfazed. She will still be reactive – it's unrealistic to expect her to change in one day – but all these little experiences will eventually lead to her being completely calm and comfortable around people.

You might not be a fan of dogs, but we can all learn something from a reactive dog like Max. Little by little does the trick. It's about putting her in uncomfortable situations, teaching her that she's safe and that there's nothing to worry about without taking her above her threshold. Eventually, as her brain is rewiring to be less scared and more confident, she'll step through the fear. Everybody's threshold is different. I know what Max's threshold is through experience and seeing what her triggers are and how far I can push her. I also know what my threshold is for how far I can go with certain things. You'll have to figure out what yours is and start pushing those limits.

'Jump in the deep end'

I've already spoken about getting dumped two days after moving to a city that I had only ever visited once before. I was

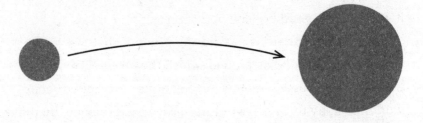

in a brand-new place with no friends. I was the oldest person on my undergraduate course as a mature student of twenty-six years old. Uncomfortable and living in the unknown was an understatement. My brain was forced to make new plastic changes in quite a drastic time frame. Imagine if you were dragged back in time to when there were no mobile phones and electricity; your brain would have to adjust very quickly to survive.

This is another way of tackling self-sabotage and fear. It's like shaving your hair, applying for a PhD at the age of sixty-five, selling all your stuff and going travelling. These are all very extreme ways of getting over fear. This route is not for everyone, but it's a route that is often taken nonetheless.

This route can also be a little less dramatic. I remember when I first went into public speaking. I received an email from a finance company asking me if I ever did any public speaking. This was just after the pandemic, when I had largely been giving talks online, so I said, 'Yes, it's mostly been online webinars, hence why I haven't got a major portfolio.' I jumped straight in. I was always good at giving presentations and I believe myself to be quite a charismatic person so I thought, *Let's go for it.* You could argue that I stretched the truth but, really, I saw an opportunity and I went for it. If I didn't say yes then, when would I have given my first public talk? I was quite nervous, but I used all the tools I knew about self-regulation, namely the physiological sigh, and I practised my talk until I felt confident about delivering my message. It was a success and I had great feedback, and thereafter I started getting more and more speaking opportunities. So I didn't shave my head and leave the country, but I still took a pretty big leap into the unknown, forcing myself out of my comfort zone.

Research[55] shows that we're more likely to choose immediate rewards, something that's a certainty now, rather than something that is unsure in the future. If the immediate reward is staying safe in your comfort zone, even if you know that stepping out of it may solve your problems, you may still choose to stay where it's safe and known. If Max could, she would never, ever go where people or other dogs are; she'd never leave the house actually. But that's an unrealistic way of living. We need to choose how we approach the fear and we have a few options. Little by little does the trick or you can jump in the deep end. Every situation is different and we can approach each one individually.

How are you going to approach your fears? I hope that it isn't by continuing to sabotage yourself. I hope that you start stepping out of your comfort zone, with all the tools and tips I've given you, and that you start to build better connections in your brain based on new experiences that challenge you to see things differently. If you were looking for a sign, for permission or for a hand to hold yours, this is it. This book is it.

Phase 3

Boost the Positive

More life: admit that you want more for yourself. Keep pressing on. You deserve more for your life. Fight for it.

- Use neuroscience to increase mental resilience
- The neuroscience behind a growth mindset
- Your muscles communicate directly with your brain
- Sleep is your number-one optimization tool
- Dopamine – your happiness is now
- Build self-trust and confidence

USING NEUROSCIENCE TO BUILD MENTAL RESILIENCE

STUDIES SHOW THAT WE CAN CHANGE OUR NEUROBIOLOGY TO INCREASE OUR THRESHOLD FOR STRESS

Increase Mental Resilience

According to the APA, resilience is 'the process of adapting well in the face of adversity, trauma, tragedy, threats or significant sources of stress'. Can we work on improving our resilience? Yes. Stress resilience can be improved by various strategies and interventions. Some individuals may naturally possess higher levels of resilience, but it's a dynamic trait that anyone can strengthen over time.

At the start of Phase 1 we discussed the function of chronic stress and the detrimental effects it can have on the body, but I also touched on the importance of acute stress in everyday life. Stress serves many purposes; for example, voluntary stress (exercise, breathwork, cold-water exposure, saunas, etc.) is responsible for making our immune system stronger. Contrary to popular belief, rather than weakening our immune system, acute stress, in fact, increases our resilience and ability to cope with future stress. Balance needs to be struck between going into a sympathetic state of stress – where your body is increasing adrenaline and cortisol to get you in a state of fight or flight – and then recovering from it. Think of it this way: if you're an athlete, you need to strike a balance between pushing yourself and training consistently, but not so much as to impact your recovery and risk getting injured.

In this chapter we're going to focus on how to improve mental resilience by increasing our ability to withstand stress and recover from it.

Voluntary stress

Through actively choosing to exercise and engage in other activities such as breathwork, voluntary stress can have beneficial effects on our brain, body and mental health. There is extensive research that shows that our mindset towards stress can also impact how we physically react to it, something we touched on in Phase 1.

In another study[6] conducted by Dr Alia Crum, one of the leading scientists researching mindsets, participants were divided into two groups. Group one was shown a video telling them the benefits of stress. This video told participants that stress enhances performance and fuels the brain and body with oxygen, increasing their energy and improving their performance, productivity, focus and decision-making abilities. The second group was told that stress is debilitating, that it can diminish drive and hinder performance. It explained that stress inhibits one's ability to think straight and make decisions, and that even trained professionals crumble under stress, then showed them a video of basketball players losing their cool on the court. The results were incredible. The group who were told that stress was positive improved their work performance and reported fewer negative health symptoms.

Similar research[5] showed that individuals who were told that stress was good for them had a positive physiological response to stress. During the experiment, their heart efficiency improved, meaning their body's ability to pump blood and deliver oxygen throughout the body was improved and used less energy. In fact, their blood pressure came down during the stressful event in comparison to those who had not been told about the benefits of stress.

Reappraising stress is a valuable tool in helping us navigate life. Understanding the difference between acute stress and chronic stress can aid us in ensuring that we strike a balance between being stressed and regulating from it so that we can function successfully.

ACUTE STRESS:
INTERVIEWS
EXAMS OR TESTS
PUBLIC SPEAKING
CONFLICT OR CONFRONTATION

VOLUNTARY STRESS:
SAUNA
EXERCISE
BREATHWORK
COLD-WATER EXPOSURE
EDUCATIONAL PURSUITS

CHRONIC STRESS:
WORK-RELATED/FINANCIAL ISSUES
RELATIONSHIP PROBLEMS
HEALTH PROBLEMS
CAREGIVER STRESS
SOCIAL ISOLATION

Stay calm under pressure

Under usual circumstances, noradrenaline (the stress chemical that is released from the brain) and adrenaline (the stress hormone that is released from the body) are released simultaneously during a stressful situation. However, trained individuals who took part in a study[56] that looked at cold-water exposure, meditation and cyclic hyperventilation (a Tibetan breathing technique that forms part of the Vajrayana practices, a form of tantric Buddhism) had different results over a period of time. This research was inspired by the Dutch motivational speaker and extreme athlete Wim Hof, who has popularized cold-water exposure and breathing techniques, and who uses these practices to educate people on the health benefits of deliberate stress. In the study researchers saw that over two weeks when the participants' adrenaline was high, brain noradrenaline was lower than the controls. In layman's terms, they were able to stay mentally calm while stress hormones were rampant in their bodies. Let's take a boxer in a boxing ring. Boxers' bodies are filled with adrenaline, yet they are able, to a certain extent, to stay calm mentally.

Top-down regulation is the brain's ability to control emotional and physiological responses to stress. Recent research[57] is starting to show us that effective top-down regulation during stressful events may contribute to increased stress resilience. When we experience stress, our brain's prefrontal cortex, responsible for reason and logic, can modulate the activity of other regions involved in stress, such as the amygdala, linked to fear and emotional processing, and the hypothalamus, which can initiate the stress response. Through this top-down regulation, the prefrontal cortex can dampen excessive stress responses and promote more adaptive coping strategies.

Here is the caveat. I told you back in Chapter 2 that the brain is designed to run, fight or freeze, and not listen, in a stressful situation, and that you can't reason with yourself when you are stressed. But this is true in moments when you're already stressed and perhaps dealing with a situation that you're un-prepared for. If you're a stock trader and are used to things going awry, when the market suddenly drops and you're in a panic you're probably able to reason with yourself because you've experienced this before. And when we deliberately choose stress, we have control over the situation and can actively keep calm through top-down regulation even before our body's stress response is triggered. For example, say you're someone who runs regularly. When you're mid-run, your body is filled with adrenaline, but your mind is in a state of calm and flow. Top-down regulation allows you to cognitively process in-formation and influence lower-level functions. For example, you could hype yourself up to run a little quicker or you could soothe yourself to stay calm under pressure.

The underlying theme here is that training yourself to be under stress, in situations where you are in control, can increase your mental threshold for it. The first time you go for a run, you may have feelings of anxiety, worry and stress. Over time it'll get easier and your brain will maintain calmness. Your body's adrenaline hasn't subsided, because it's a necessity, but your brain's noradrenaline has. This means that the link between the two systems has been weakened and your body can be under stress while your mind remains calm. And this experience can be transferred to your ability to withstand sudden stress from your boss, a conflict or something going wrong in your life.

Of course, this concept is multifaceted, and we can never fully separate the two systems so that they don't affect one another. But we can improve our resilience to stress by choosing to put ourselves in stressful conditions while knowing that we're safe.

Mindfulness meditation

Mindfulness meditation training has also been shown to improve our resilience and ability to deal with physical stressors. Studies[58] have shown that there is a greater drop in stress hormones and inflammatory molecules in the body during stressful tasks in individuals who have adopted mindfulness training (meditation). Several studies[57,59] have suggested that having strong top-down regulation can improve our stress resilience and help us recover from stressful events. This can be developed through meditation, CBT and other practices such as breathing techniques. Meditation enhances top-down regulation by improving attention control, emotional regulation and self-awareness through strengthening the prefrontal cortex and improving overall cognitive function.

However, it is essential to note that stress resilience is a complex trait, influenced by various genetic, environmental and individual factors. While top-down regulation is likely to play a role, it is not the sole determinant of stress resilience. Other factors, such as social support, coping skills and personal experiences, also contribute to an individual's ability to navigate stressful situations.

Love invents us

There is also evidence that psychological resilience can be highly associated with your environment and the support network you have around you – some of those being family, school and peer support, as well as self-regulation and self-care, which we'll look at in more detail later. Social connection and support are associated with more resilience because social connections boost resilience through emotional support, a sense of belonging and shared coping resources. In times of adversity, a supportive network offers comfort and understanding, acting as a buffer against stress. Belonging to a social group fosters identity and acceptance, reducing the feelings of isolation that can hinder resilience. Trusted friends and family give a different perspective to our issues; they also provide valuable advice which can enrich problem-solving strategies. But more importantly they can help us create a narrative around our problems in life.

I had a situation recently where a friend of mine described an issue she was having with her boyfriend, and as an outsider looking in, I was able to give her a more balanced outlook on the situation. This helped her see things from a different perspective, create a more positive narrative around the problem and ultimately find a suitable solution and approach. To add to this, positive social connections contribute to our physical well-being by helping to lower our stress hormone levels and improve our immune function. Oxytocin and serotonin release can aid in a sense of catharsis. These health benefits enhance our capacity to rebound from challenges. The practical and emotional assistance we get from social networks aid our emotional coping.

You might not have anyone in your immediate family or friend circle. I have come to learn this from some of my Instagram followers, predominantly from men, who say they don't have supportive friends. There are many charities, support groups and online pen pals that can help you access the emotional support you need from your peers – perhaps you can join a sports group, a book club or a running club. I think you'll come to realize that most people need someone they can talk to. Help is out there for you; you just have to find it.

Dr Bruce Perry is a child psychiatrist and he has seen some of the world's most horrific cases when it comes to childhood neglect and abuse. But in his book *The Boy Who Was Raised as a Dog*, he concluded that no matter what any child had gone through, the ones who thrived later in life were those who had support – either from their family or from the foster homes and support groups they were in. You may not have received support as a child, but you can learn to foster relationships later in life that are nurturing.

Humans need one another: we need connection and we need to feel like we're not alone. Reach out to your friends or build a solid network of support through extracurricular activities or therapy/group therapy. Love invents us. For many years we have been taught that 'nobody will love you unless you love yourself' but the reality is that you cannot learn to love yourself unless you have been loved by others before. This doesn't mean that you need to rely on others for their validation and not put in your own work, but consider the fact that brain maps are areas in the brain that encode various modalities, and they form through somatosensory input. Therefore the capacity to love cannot be built in isolation, and it has been proven by Dr Perry that children who have endured trauma will recover and thrive if they are surrounded by people who love them. For those who are intro-verted or neurodivergent who don't thrive on social interaction; research[60] shows that the quality of social connection is more important that the frequency. We can spend about 75% of our time alone. This means that even just having a few meaningful connections can help us feel connected and supported.

Neuroplasticity and resilience

Teaching our brain that we can bounce back from setbacks means that when we inevitably 'fail', by means of experience-dependent plasticity we will be more resilient and equipped to deal with the pitfalls. Our brains are designed to learn from failure and to add to the database of experiences of how to and how not to respond in future situations. Failure is necessary for future success because it teaches our brain about the world and about ourselves. By harnessing the power of neuroplasticity and seeing failure as a positive learning outcome, we can build a resilient mindset, which enables us to navigate life's challenges with greater fortitude, maintain mental well-being and ultimately lead fulfilling and meaningful lives. This empowers us to constantly evolve and adapt, highlighting the incredible potential for growth and recovery that lies within the human brain.

REWIRING ESSENTIALS – USE NEUROSCIENCE TO INCREASE MENTAL RESILIENCE

- Some individuals may naturally possess higher levels of resilience, but it's a dynamic trait that anyone can strengthen over time.
- Voluntarily putting yourself through stressful situations in a controlled manner will increase mental resilience by increasing your ability to withstand stress and recover from it.
- Research shows that our mindset towards stress can also impact how we physically react to it.
- Understanding the difference between acute stress and chronic stress can aid us in ensuring that we strike a balance between being stressed and regulating from it so that we can function successfully.
- Deliberately choosing stress and actively keeping calm teaches your body to stay calm under pressure.
- Our brains are designed to learn from failure, and to add to the database of experiences of how to and how not to respond in future situations. Failure is necessary for future success because it teaches our brain about the world and about ourselves.
- By harnessing the power of neuroplasticity and seeing failure as a positive learning outcome, we can build a resilient mindset.

FIXED VS GROWTH MINDSETS

FIXED	GROWTH
AVOIDS CHALLENGES	EMBRACES CHALLENGES
GIVES UP EASILY	PERSISTS
SEES EFFORT AS WORTHLESS	SEES EFFORT AS MASTERY
CAN'T TAKE CONSTRUCTIVE CRITICISM	LEARNS FROM CONSTRUCTIVE CRITICISM
FEELS THREATENED BY OTHERS' SUCCESS	IS INSPIRED BY OTHERS' SUCCESS

Growth Mindset

'In a growth mindset, people believe that their most basic abilities can be developed through dedication and hard work – brains and talent are just the starting point. This view creates a love of learning and a resilience that is essential for great accomplishment' – Dr Carol Dweck

I told you about my friend Samantha, the dancer. Her sister Martha was a high-achieving student; she was at the top of her class and her parents always raved about how intelligent she was. She got into a top graduate-scheme programme at a big digital marketing firm after graduating *cum laude* in her Master of Fine Arts degree. However, she struggled at work and was deeply insecure. She found it hard to understand why she couldn't cope with life as an adult when she was so intelligent. The thing with Martha is that even though her IQ was so high and she was undoubtedly very smart, her mindset was fixed. She was afraid of putting herself out there at work when it came to problem solving because she was afraid of failure, and, even worse, she was afraid that people would doubt her intelligence.

I suggested to Martha that she see a therapist to help her adjust to her new environment, which she did. Her therapist helped her work on changing her mindset. Over time she learned to rewire her behaviour and beliefs to be more open to failure, while breaking down a lot of her previously held ideas about herself so that she could stop attaching her identity to the outcome of a problem or her performance at work. Martha had to learn that her sense of identity was so attached to being intelligent and that anything that suggested otherwise felt like an attack on her as a person.

If this sounds like you, I want you to know that this can be changed. Many people attach their identity to their success and outcomes, so much so that it holds them back from truly achieving anything. They become perfectionists and they avoid difficult challenges due to a fear of failure because they take this failure personally; so in an attempt to avoid diminishing their self-worth, they remain stuck and avoid confronting problems. Unfortunately the world doesn't work like that. Martha couldn't avoid the problems at work but she couldn't learn from them either until she sought help.

Fixed mindset vs growth mindset

Dr Carol Dweck is one of the leading researchers in the field of mindset and motivation. Her research has highlighted some valuable lessons on how we tend to stay within a fixed mindset when we attach performance and outcomes to our identity. In one experiment,[61] teachers gave their students feedback that linked their intelligence to their identity and gave comments such as, 'You're so smart!' Later, when these students were asked to select problems to solve, the study found they favoured less challenging ones. They chose problems that would reinforce their belief about being smart and allow them to continue to demonstrate their good performance. The study also found that students would lie about their scores to amplify their level of intelligence. On the other hand, schoolchildren who were given feedback based on their effort, given comments such as, 'You must have worked hard at these problems,' chose problems later in the study that promised increased learning. These children were less concerned with the other children's scores and preferred to receive strategy-related information on how to solve future problems.

These studies and others that came after[62,63] show us that when we attach our identity to the outcome of a particular action, like Martha did, we attribute our motivation and performance to who we are as a person. This means that when we are faced with challenges, we are more likely to choose an easier route because we don't want to reinforce the belief that we are incapable of achieving something. When we adopt a growth mindset, we learn to attach the outcome to the effort we put into a task and learn to understand that challenges can lead to learning opportunities because we recognize that our capabilities are not fixed; they can be improved with effort. Our identity isn't attached to the outcome, which means we can separate ourselves from it.

A growth mindset is the belief that our abilities and intelligence can be developed through dedication, effort and learning.[64] It allows us to embrace challenges and overcome them. People with a growth mindset understand that their abilities can be developed. And those with a fixed mindset may find problems catastrophic and feel as if they are being judged.

Learning is vital

The first major underpinning of a growth mindset is that people with this mindset understand that *learning is a valuable opportunity in the face of adversity*. When people believe that they can improve and grow from failure and setbacks, they are more likely to engage in challenging tasks and persist through difficulty. When people know and understand that the brain is malleable and are willing to adapt to circumstance, they are more likely to persist in the face of obstacles. This perseverance can enhance pathways in the brain that are associated with learning, which strengthens the notion that learning is a dynamic process that's forever evolving.

The brain circuitry in those with a growth mindset shows that they are more receptive to correction and are able to neutralize negative feedback. Growth mindset has been associated with the striatum and basal ganglia, both key areas responsible for reward-related learning.[65] Children who underwent growth-mindset training showed an increase in activity in those areas over the course of four weeks when they engaged in difficult maths problems. Additionally, the schoolchildren from the previous study with a growth mindset were able to perform with higher accuracy during tests even after making mistakes. Not only that, but students were also able to remember and make sense of their mistakes, thus self-correcting during the learning process, boosting autonomy and confidence.

Identity does not equal performance

The second major underpinning of a growth mindset is *detaching our identity from the outcome*. Martha attached her work performance to her personality, which meant that when she started making mistakes and struggling at work, she internalized this as a personal failure, as if there was something wrong with her. This promotes a fear of failure because it is attributed to the individual.

When people detach themselves from the outcome of a goal or a challenge, and learn to understand that fear of failure is, in fact, an opportunity for growth, they can fail without undermining their identity. Djokovic says that even if he loses, he has no regrets. We should all try to look at life in a similar way. Because together these two features of a growth mindset can help individuals excel in life and foster a journey of lifelong learning without the fear of failure which inevitably holds us back.

If you're someone who currently has a fixed mindset, this is your reminder that the brain is plastic. We have been breaking down preconceived ideas and beliefs since the beginning of this book. Let's take a look at what our biases are driving and how they're affecting us.

ACTIVITY

What story are you replaying in your mind that keeps you fearful of outcomes?

_____

_____

_____

What labels have you used to describe yourself that keep you in a fixed mindset?

_____

_____

_____

Do you think that your challenges in life really dictate who you are as a person?

_____

_____

_____

Can you change those beliefs to see that there is a learning opportunity in every failure and that it doesn't reflect on your identity?

_____

_____

_____

Learn to detach your identity from the outcome. If you believe that your qualities are fixed, you will stay stuck, trying to prove yourself right every time. You will avoid mistakes and you'll reinforce negative self-beliefs when you fail.

Our personal qualities are not fixed; intelligence is not fixed. If you're not very knowledgeable on a topic or you're not very studious, you can learn to get better through practice. By believing that our qualities can be developed and changed, we start to change our entire view on things. Remember what we learned in Phase 1 about confirmation bias, how our beliefs drive our thoughts and actions, and how they affect one another. By believing that your capabilities and qualities can change through effort, you start to shift the way you carry yourself and the way you see failure; and, just like the schoolchildren in the study, you may become less bothered by what those around you are doing and more committed to your own growth and learning.

I want you to reflect on the last failure that you attached to your identity. Can you reframe it to see that it had nothing to do with who you are as a person?

HOW YOUR BRAIN MAY START REWIRING ITSELF
WHEN YOU CHANGE YOUR MINDSET

Reward

Regions of the brain associated with effort and persistence are more active and exerting effort becomes a rewarding task as you begin to see things as opportunities to learn and improve.

Positive reinforcement

When you encounter setbacks, your brain will begin to respond differently to errors and mistakes. Reward centres are activated during obstacles, which can further reinforce encouragement and growth. This will positively reinforce your belief about growth mindsets.

Reduced fear of failure

Regions of the brain associated with fear and anxiety will be reduced. This will make you less fearful of failure and more likely to become more confident.

Knowing this about growth mindsets, coupled with what you now know about stress mindsets, we can start to see how we can change our view on life to propel us towards success.

One of the things that really grounds me when I am going through a hard time is remembering that you learn a lot about yourself when you're in the darkness. Like my best friend would say, 'We transform in the tight, claustrophobic cracks in between deciding to undertake something and actually getting somewhere. In those dark and "stanky" cracks.'

I agree with her. I think the magic happens for us in places of discomfort. And every time I think I have been buried, I have actually been planted. And when I get ripped out of the ground again, I come to learn that it's a harvest. Life will pick you up and put you back down. It's not fair; it never has been and it never will be. I have been humbled by it so many times. Now the thing is, you can either choose to use downfalls as learning opportunities, or you can choose to live offended. It's a tough choice, but if one thing is certain it is that you can't choose what life is going to serve you.

Sometimes the world knocks you around a little bit to remind you that there's something worth fighting for.

Keep fighting for your place in this world. Keep fighting for your narrative, your story, your place in the driver's seat.

REWIRING ESSENTIALS – THE NEUROSCIENCE BEHIND A GROWTH MINDSET

- Fixed mindset: avoids challenges, gives up easy, can't handle constructive criticism.
- Growth mindset: embraces challenges, learns from constructive criticism, is inspired by success.
- Studies show that when we attach our identity to the outcome of a particular action, we attribute our motivation and performance to who we are as a person.
- If we do this, then when we fail, we consider ourselves to be failures.
- When we adopt a growth mindset, we start to understand that challenges can lead to learning opportunities.
- Our capabilities are not fixed; they can be improved with effort.
- Performance does not equal identity.
- Learning is vital.
- Your brain can rewire itself towards seeing challenges and failures as growth experiences.

THE MUSCLE-BRAIN AXIS

HOW EXERCISE PROTECTS THE BRAIN FROM DEPRESSION, ANXIETY AND NEURODEGENERATION.

Your Muscles Communicate Directly With Your Brain

Some of the most common feedback I get from my Instagram followers is how much they appreciate the scientific grounding that I give to exercise. I think it's because so many people are bored of hearing about the benefits of exercise without really understanding why it's so important. Many people are also afraid of exercise, or they see it as a form of punishment for eating too much or as a dreadful activity that one must partake in to lose weight. Exercise has got a bad rap over the years. From the diet culture sprawled across the media to the bodies of unrealistic fitness influencers, exercise has been made confusing and unattainable . . . and somewhere along the way we forgot about the real benefits.

This chapter breaks down the science behind why exercise is beneficial from a neurophysiological perspective and what specific molecules are associated with these benefits. I get so many comments and messages from people to say that I have inspired them to start exercising because I have taken away the shame and confusion surrounding the topic and empowered people with scientific facts and knowledge that have helped them change their view on it.

Benefits of exercise on our brain health

Increased neuroplasticity

Grey matter is the part of the brain that is composed of neuronal cell bodies, dendrites and synapses that communicate with other neurons close by. It appears grey due to a lack of myelin.

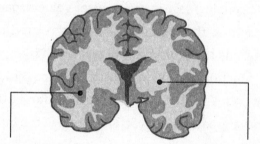

Grey matter
- Contains most of the brain's neuronal cell bodies
- Fully develops once a person reaches their 20s
- Involved in information processing
- Involved in higher-order thinking: perception, memory, learning and decision-making
- Able to rewire

White matter
- Made up of bundles which connect various grey matter areas
- Develops throughout the 20s
- Interprets sensory information from various parts of the body
- Acts as a communication network, facilitates the transfer of information across various brain areas
- Able to rewire

Myelin is a fatty sheath that covers neurons and acts as an insulating layer, which helps conduct information to the more distant regions of the brain and spinal cord. White matter appears white because of these fatty sheaths. Grey matter is responsible for processing information, cognition, sensory perception and muscle control. Neurodegenerative diseases often involve the loss of grey matter. Conditions like Alzheimer's disease, Parkinson's disease and Huntington's disease are characterized by the progressive degeneration of neurons and the associated loss of brain tissue, including grey matter. This loss can lead to cognitive decline, memory loss, motor problems and other symptoms associated with these disorders.

Research[66] shows that regular physical activity is associated with a significant increase in grey matter volume, particularly in the hippocampus, the region of the brain related to memory and learning, and the prefrontal cortex, which is responsible for executive function, decision-making and cognitive control. Adolescents with higher levels of aerobic fitness have been shown to have greater grey matter volume in the hippocampus compared to those with poorer fitness levels. These adolescents also score better on cognitive tests.

Physical activity appears to be a promising method for influencing grey matter volume in late adulthood as well by promoting proteins in the brain that correlate with cognitive integrity.

Inflammation and the body

Inflammation is a necessary part of the body's healing process. When you get sick or have a cut or infection, inflammation increases in the body to repair the areas that need attention. This inflammation usually subsides until needed again, but when we are chronically stressed, our bodies become inflamed and are in a constant state of believing they need repairing. And when the 'threat' or stress doesn't go away, the inflammation becomes chronic.

Chronic inflammation can be caused by a variety of factors, including persistent infections, autoimmune diseases, chronic stress, unhealthy lifestyle choices and environmental stressors such as pollution. It can attack the whole body and even cross into the brain through something called the blood–brain barrier (BBB). The BBB is tightly regulated and its function is to regulate substances between the blood and the brain, helping maintain its delicate environment and protect it from harm. This is important information that will be useful when we talk about depressive symptoms.

Inflammation in the brain can lead to an array of issues including neurodegeneration, depression and potentially anxiety.[67] Of course these issues are multifaceted and may require a multimodal approach, such as therapy in conjunction with stress reduction and anti-inflammatory interventions. But from a neurological standpoint this is what is happening on a biological level. Remember, if your brain health is your hardware and your mental health is your software, it would be hard to make any software updates on a piece of hardware that isn't working optimally.

Kynurenine and tryptophan

We know that serotonin, one of the brain's feel-good neuro-transmitters, helps modulate mood and other brain functions such as sleep and sexual desire, which can indirectly affect our mood. Serotonin is made from an essential amino acid called tryptophan, which we get from foods such as nuts and seeds, tofu, chicken, turkey, cheese and many fruits; tryptophan converts into serotonin and then melatonin. We call this the tryptophan pathway, and within this pathway is a molecule called kynurenine, which plays a significant role in the body's stress response.

When we experience a significant amount of stress, inflammation drives this pathway to create more kynurenine instead of serotonin.[68] Kynurenine levels in the brain are delicately balanced to provide benefits, but when there is an imbalance – in this case, an abundance – it has detrimental effects on the brain and body. This diversion leads to decreased serotonin production, which can contribute to depressive symptoms and low mood.

Additionally, kynurenine is further metabolized into quinolinic acid, which is neurotoxic for the brain.[69] This inflammatory mechanism can also exacerbate feelings of anxiety and worry, which overlap with stress and depression. So we can see how a pathway that is supposed to create serotonin and work in our favour can be heavily disrupted in the presence of stress and create a downstream of negative effects on the brain which contribute to depression through a variety of mechanisms.

1. Less tryptophan is available for serotonin production.
2. Quinolinic acid is neurotoxic, which can lead to neurodegeneration.
3. Kynurenic acid could also contribute to depressive symptoms by increasing inflammation in the brain.
4. Inflammation may contribute to feelings of anxiety and worry.

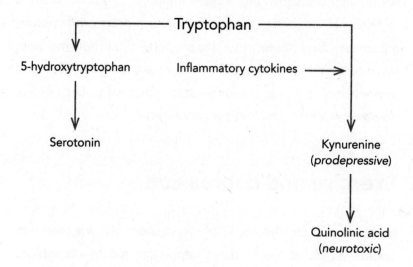

This is where exercise comes into the picture

When we exercise or perform any physical activity or movement, our brains signal our muscles to contract and relax. The mechanistic movement of you contracting and relaxing your muscles releases enzymes that break down kynurenine before it can reach the brain, cross the blood–brain barrier and cause negative effects. Additionally, this mechanistic movement also releases myokines.[70] Myokines are muscle-based proteins that have a wide variety of functions.

These myokines communicate directly with the brain through something called the muscle-brain axis. This axis highlights the profound connection between physical activity and mental health because your muscles have a direct line of communication to your brain; regular exercise plays a crucial role in maintaining a healthy and resilient mind.

Myokines aid in alleviating depressive symptoms and improving anxiety, have neuroprotective properties[71] and help maintain synaptic integrity, ensuring our neurons are firing correctly and information is being communicated effectively. Let's break down some of the myokines in more detail.

Exercise and depression

Here we discuss the topic of depression, but we need to acknowledge that symptoms of depression are on a spectrum. Everybody's experience and where they are on the spectrum is different and may require additional interventions to help alleviate symptoms. But this is what we know when it comes to the neuroscience of depression in regard to exercise interventions.

BDNF
BDNF is a molecule/protein that supports the growth, survival and function of neurons in the brain and peripheral nervous system. BDNF plays a crucial role in promoting neuroplasticity[72]

and ensuring that synapses stay strong and healthy. Several studies[72-74] have shown an association between lower levels of BDNF and the development or severity of depressive symptoms. This is because depression is associated with impaired neuroplasticity, which may contribute to the reduced ability to adapt and cope with stress and other negative emotions.

IGF-1

Another interesting myokine is insulin growth-like factor 1 (IGF-1). IGF-1 is a hormone that plays a crucial role in growth and development, particularly during childhood and adolescence. It also has neuroprotective effects in the brain like BDNF, meaning it supports the survival, growth and function of nerve cells. IGF-1 is usually released from the liver, and an abundance of this hormone can be associated with a wide variety of issues. However, this is the interesting part: IGF-1 acts on different pathways and has different positive outcomes when released as a myokine[75] from the muscle. So a molecule can have different effects depending on the context in which it has been released.

IGF-1 is predominantly released during weight training, and if you're somebody who has never lifted weights before, it's really important to understand that overloading our muscles progressively, what we call progressive overload, is what's going to help them stay strong and provide a wide range of physical and mental benefits. There is a workout guide at the end of this chapter which will outline the key requirements for exercise per week to gain the most benefit for your brain and mental health.

Exercise and anxiety

We learned about anxiety in Phase 1 and how to manage some of those feelings of worry and apprehension. Regular exercise is a powerful ally in managing anxiety. Physical activity triggers the release of endorphins, neurotransmitters that act as natural mood-lifters, and more recently we've come to learn about the effects of myokines on anxiety through the muscle-brain axis.

IGF-1

Studies[76] show that reduced levels of circulating IGF-1 are associated with the increased presence of anxious symptoms. IGF-1 has also been shown to play a key role in synaptic plasticity, the ability for neurons to modify their connections, whereby it favours fear extinction in memory by weakening the synapses. Fear extinction is like rewriting the brain's 'scary' script, helping it learn that once-threatening things are now safe. Basically IGF-1 weakens synapses in the hippocampus[77] that hold memories about certain situations and diminishes the links between those memories and the fear attached to them.

EMDR and walking

One of the best techniques in psychotherapy is EMDR therapy. EMDR is used to treat individuals who have experienced traumatic events or distressing life events. Research has shown

EMDR has major benefits for emotional reprocessing, but it wasn't until recently that neuroscientists discovered the mechanism that reveals why it works.

Lateral eye movements are believed to play a role in the process of fear-extinction learning because when our eyes move from side to side they engage the frontoparietal network.[18,78] This is a network of brain areas located at the front of your brain that is involved in attention, problem solving and working memory. The frontoparietal network works in tandem with the amygdala, our fear-processing centre, meaning that it competes for resources with it. In the same way that acute stress impairs your ability to think and concentrate on problem solving, the opposite is happening when you are engaging the frontal cortex: the amygdala becomes disengaged.

This is great knowledge that applies to any type of movement that puts you in forward motion. Things like walking, running and cycling mean that the world moving past you activates this frontoparietal network while reducing amygdala activity. So these types of exercise can help us make sense of any issues we're having in our lives while also processing them without the anxiety and worry that is usually attached to them. It helps us rationalize our thoughts and gives us the space to step back from our issues and see them from a different perspective. You'll find that even if you're not consciously thinking about your problems, your brain will most likely still be problem-solving subconsciously.

If you're anxious about running and cycling, start with a

small walk. The benefits of aerobic exercise are found in zone 2 of your heart rate. Zone 2 is a moderate intensity of working out where you work at a level that still allows conversation. This means that you don't need to have a total blow-out where you're sweating and struggling. A brisk walk where you can still hold a conversation is extremely beneficial for overall health. Everyone's zone 2 will be different; for some it'll be a light jog, and for others (like me) it's walking a dog or cycling on a flat road. Zone 2 is working in your aerobic zone, meaning that you should be able to sustain it for a long period of time without getting fatigued.

HOW EXERCISE ENHANCES MOOD

ONE
Exercise, predominantly aerobic exercise such as running, swimming, walking and cycling, increases blood flow to the brain, which enhances brain oxygenation and the delivery of nutrients to neurons. This stimulates the production of BDNF, which then binds to receptors and increases their integrity.

TWO
Exercise releases feel-good neurotransmitters such as dopamine, norepinephrine and serotonin, which alter your neurobiology to improve your mood. Dopamine and serotonin are brain chemicals that can support your mental health, and they even stimulate the release of BDNF. Low levels of these neurotransmitters have been linked to symptoms of depression.

THREE

Endorphins act as a natural painkiller that is created in the body. They are released during exercise and other activities such as laughing and crying. The word 'endorphin' means 'endogenous morphine' because it has morphine-like effects in the body. They are responsible for dampening pain and promoting feelings of well-being and euphoria. They weaken pain signals and create a sense of pleasure and reward.

FOUR

In previous chapters we spoke about stress resilience and how we can use exercise to improve our threshold for stress. Exercise can help the body's stress system to become more adaptable to voluntary stress so that we can better deal with unforeseen stressful events in other aspects of life.

FIVE

In Chapter 2 we learned about the importance of hobbies in regulating our stress response. We also learned in Chapter 4 that our thoughts are extremely powerful and that if our brain is still thinking about stressful events, it still perceives our body to be under threat. Exercise acts as a temporary distraction because it requires focus and concentration; this can distract us from the negative thoughts and rumination that are associated with depression, giving our brains time to recover. Physical activity can help us break the cycle of negative thinking patterns by essentially 'rebooting' our thoughts.

SIX

Engaging in aerobic activity increases the mitochondria in our brain and body. The mitochondria is the powerhouse of the cell, responsible for producing energy in every single cell of our body. This plays a crucial role in cellular health and function. Mitochondria are responsible for generating adenosine triphosphate (ATP), the most basic energy form in our system – think of it like energy currency. Adequate ATP production is important for maintaining optimal brain function and has been linked to improved mood and mental clarity. Mitochondria play a role in the metabolism of neurotransmitters as well. Proper regulation of neurotransmitters, such as serotonin and dopamine, is critical for mood regulation. Dysfunction in the mitochondrial processes may disrupt neurotransmitter levels and contribute to mood disorders. Mitochondria are also involved in managing oxidative stress and inflammation in cells. Chronic oxidative stress and inflammation have been linked to various mental health conditions, including depression and anxiety. Therefore, improving mitochondrial function can potentially reduce oxidative stress and inflammation, benefiting mood.

SEVEN

Grey matter volume can be shaped and moulded through exercise. Regular physical activity, both aerobic and resistance training, has been shown to promote neuroplasticity. Aerobic exercise has been linked to increased grey matter volume in key regions associated with learning and memory, such as the hippocampus.

Exercise helps the release of myokines, which in turn support the growth and maintenance of neurons and promote brain longevity and integrity. This is beneficial to us in many ways, by aiding how we learn new things and all the way through to our ageing process, brain health and longevity.

EIGHT

Neuromotor training, which involves exercises that enhance coordination, balance, agility and proprioception, stimulates neuroplasticity. Engaging in these activities encourages the brain to create and strengthen these connections. Research[79] shows that neuromotor training may even help stave off neurodegenerative diseases by reducing the risk of cognitive decline.[80]

ACTIVITY

Can you reflect on a time when you were ruminating about a problem or perhaps even about yourself and went for a walk and felt better afterwards?

_____

_____

_____

If this doesn't resonate with you, can you find a way to incorporate this type of exercise into your week? Perhaps twenty to thirty minutes per day to clear your head.

_____

_____

Come back to this and reflect on your experience of taking up aerobic exercise to improve your mood.

_____

_____

_____

_____

Your recommended weekly regime

While some myokines are expressed more during aerobic exercise, many of them are expressed during weight and resistance training. My suggestion is to adopt an exercise regime that includes both each week.

- Approx. 150 minutes of zone 2 aerobic training – steady-paced movement
- 3–8 minutes of zone 5 training – short and challenging
- 2–3 weight training sessions – resistance training is arguably the best method of training for health and longevity
- Integrate neuromotor training where possible – balance/coordination.

Your Muscles Communicate Directly With Your Brain

Zone 1
Very light activity – warm up/cool down

Zone 2
Light activity – slow-paced jogging, walking up a flight of stairs,
lightweight low resistance

Zone 3
Moderate activity that increases aerobic endurance –
moderate jogging or cycling

Zone 4
Hard anaerobic activity – ball slams, boxing,
interval running or heavy weight-lifting

Zone 5
Extremely hard maximum-exertion activity –
sprinting or Tabata, all-out effort

REWIRING ESSENTIALS – YOUR MUSCLES COMMUNICATE DIRECTLY WITH YOUR BRAIN

- Many people are afraid of exercise, or they see it as a form of punishment for eating too much or as a dreadful activity that one must partake in to lose weight.
- Exercise is much more than that. It has a direct line of communication to your brain and elicits a wide range of benefits to your brain health.
- Exercise helps release myokines, which are muscle-based proteins that communicate with the brain through the muscle-brain axis.
- Myokines aid in alleviating depressive symptoms and improving anxiety, have neuroprotective properties and help maintain synaptic integrity, ensuring our neurons are firing correctly and information is being communicated effectively.
- Exercise has anti-inflammatory properties which protect the brain from neurodegeneration.
- Exercise can boost neuroplasticity through the release of BDNF, a protein that protects neurons and synapses.
- Exercise enhances mood in various ways, predominantly through myokine release; endorphins and endocannabinoids, which are endogenous painkillers that make you feel good; and increasing serotonin, norepinephrine, dopamine and oxytocin – feel-good neurotransmitters.
- Exercise also increases mitochondrial density, which improves the energy consumption of every single cell in your brain and body, making you more efficient at utilizing energy – helping you feel more energized.

SLEEP IS MUCH MORE THAN YOU THINK:

REM SLEEP IS IMPORTANT IN REMOVING THE EMOTIONAL LOAD OF TRAUMATIC EXPERIENCES AND INTRUSIVE THOUGHTS.

Sleep Is Your Number-One Optimization Tool

I know. Eye-roll. That's the usual response I get from people when I tell them how important sleep is. I was having a chat with my friend's father, who doesn't understand much about physiology, and he giggled when I told him that the solution to his stress and weight gain is to sleep more.

'Lack of sleep doesn't make you gain weight,' he said.

'Oh, but it does, Michael,' I said.

You see, when you don't fully understand what's going on at a physiological level, it's hard to put two and two together. How could sleep possibly affect weight gain and other aspects of your life like your immunity and your health?

I remember watching *The Intern*, starring Robert De Niro. In the film he plays a pensioner looking for a purpose, and he decides to apply for a job as a senior intern at a fast-growing online fashion brand. The owner and CEO of the brand, played by Anne Hathaway, is so highly committed to her work that her marriage is failing. This is putting her under a lot of pressure to hire someone else for the job and therefore she isn't sleeping a lot. In one of the scenes, Hathaway's mother says that women who sleep less than seven hours a night are 38% more likely to experience major weight gain.

I remember hearing that in 2015 when the film came out and feeling like it was click-bait information given by wealthy movie mothers that fed in to diet culture (picture Regina George's mother in *Mean Girls*). But actually it turns out they were giving some factual information. While the actual percentage of weight gain may not be correct, there is a link between sleep deprivation and the risk of weight gain. I didn't know this at the time but it clicked for me when I learned about sleep at university and read Matthew Walker's bestselling book *Why We Sleep*. Now I'm not here to talk about weight loss per se, but I do want to highlight the importance of sleep for our overall health, especially how it affects our ability to create behavioural changes.

Sleep is arguably the most important step in creating new habits and making changes. It's when all our memories and new acquired information become consolidated and encoded in the brain's memory centres. When we're awake, the brain is actively processing and acquiring new information. Initially these memories are fragile and susceptible to decay. When we sleep, the brain stabilizes these memories by forming new synaptic connections in our memory centres so that we can later recall the information. This solidifies new memories and the information you've acquired during the day. Additionally, you may have experience with brain fog and an inability to think straight when you've slept badly. I'm well acquainted with that feeling, and anecdotally it seems most people are. This also impacts how we make new memories the subsequent day.

Sleep is important for so many biological functions, including our ability to rewire our brains. In Part 1 we spoke about stress and how the brain's ability to make plastic changes is compromised when in low-power mode. Lack of sleep puts us in that state. Sleep is an indispensable physiological process that profoundly impacts the human body and mind. Beyond memory consolidation, it also plays a vital role in maintaining physical health, cognitive function, emotional stability and overall quality of life.

This chapter is going to discuss sleep in relation to neuroplasticity but I also want to highlight some other important biological functions that are affected by lack of sleep. To understand sleep, let's break down what our brains should be doing at night.

Stage

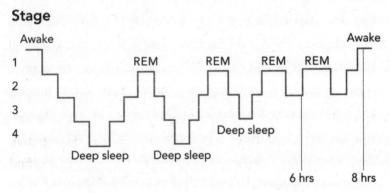

Deep sleep – growth hormone release
REM – testosterone

When we sleep, we cycle through two phases of sleep: rapid eye movement (REM) and non-REM sleep, with each cycle lasting approximately ninety minutes. Early in the night, we go through two stages of deep sleep, the slowest-wave sleep that is the most restorative for our brains and bodies. If your sleep is well curated, you can reach deep-sleep stages three times in the night. Typically you'll descend into a deep sleep within an hour of falling asleep, with each cycle getting progressively shorter as the night progresses. This stage is critical for restorative sleep, allowing the body to repair and grow and for new memories to be consolidated. It's also responsible for cleaning out toxins in the brain through something called the glymphatic system. This system ensures the efficient elimination of toxic metabolites from the central nervous system; think of it as a waste-clearance system for the brain. When this system isn't functioning properly, it can lead to neuronal loss, inflammation and potentially dementia.

There is evidence to suggest that honing your deep sleep can contribute to insightful thinking and improve memory and learning. This is all really important information for those of us on this journey to rewiring our narratives. You may even find that when you're learning something new, such as a new language, you're more tired and have less available cognitive energy. You start forgetting things, misplace your keys and find yourself spacing out.

Sleep and processing intrusive thoughts

REM sleep is crucial for emotional processing and memory consolidation. During REM sleep the brain is highly active and dreaming occurs. This stage of sleep is associated with emotional regulation, memory consolidation and the processing of emotional experiences. Studies[81] have shown that REM sleep plays a significant role in integrating and processing emotional memories; it helps to remove the emotional load associated with traumatic experiences and intrusive thoughts. This process involves reorganizing and restructuring memories, which can contribute to a reduction in emotional reactivity and the distress associated with traumatic events. Therapeutic approaches such as EMDR have been developed to mimic some aspects of REM sleep to help individuals process and manage traumatic memories. EMDR involves guided eye movements and is believed to assist in reducing the emotional impact of such memories.

During REM sleep our norepinephrine levels are relatively low compared to wakefulness. Norepinephrine is responsible for alertness and arousal. So when norepinephrine levels are low, this helps with emotional processing and the removal of the emotional load associated with intrusive thoughts. Therefore the decrease in norepinephrine during REM sleep may contribute to the brain's ability to process emotions and memories without triggering a person's emotional centres.

Testosterone

Both men and women have testosterone; women have it in lower quantities, but it's a vitally important hormone for both. The majority of testosterone release in men occurs during REM sleep, and while the data on women and testosterone is limited, we do know there is evidence to suggest that low testosterone levels in women are linked to lower levels of sleep efficiency. It's a bidirectional relationship, where lack of sleep lowers your testosterone and low testosterone impairs sleep duration and quality.

Testosterone is mostly released during REM sleep, so sleeping for just six hours means you're losing an entire cycle of REM, which also means you're losing an entire cycle of testosterone release. A study[82] in healthy adults found that if men slept five hours a night for eight nights, their testosterone levels dropped by 10–15%. This is a drastic amount; to put it into perspective, normal ageing decreases testosterone by 1–2% per year.

Testosterone is a key regulator of cognitive function with neuroprotective properties. It impacts our ability to fall asleep and is also responsible for muscle growth and bone density, all of which are forfeited if we don't get enough sleep.

How is this relevant to rewiring?

Testosterone heavily influences synaptic plasticity, particularly in the areas of the brain responsible for learning and memory.

It also modulates the bonding of new connections in the brain, ensuring that what you're repeating is sticking.

Testosterone is linked to improved cognitive function and attention. We spoke about the importance of attention towards what we want to change. Testosterone is important in ensuring that we stick to the game plan by bringing behavioural changes to our attention every day.

Testosterone aids in cognitive functions that allow us to adapt and change in response to new experiences.

Testosterone regulates our mood and emotions. This aids emotional processing and may even play a role in stress resilience and emotional stability, helping our brains adapt to different emotional challenges.

Growth hormone

Growth hormone, on the other hand, is mostly released during deep restorative sleep. Whether you're an athlete or someone who has a stressful job, growth hormone is there to aid in all restorative functions regardless of your activity levels, including the damage caused by chronic stress, so it's not just important for bodybuilders and growing children but vital for everyone and anyone. Growth hormone regulates metabolic functions such as insulin action and blood sugar levels, and it also helps with repairing cells in the brain that support neuron function. These supporting brain cells help with cognitive function and regulate mood. Growth hormone can improve alertness and endurance, and reduce irritability; therefore getting not just a good amount of sleep but deep quality sleep will help with your overall physical, emotional and cognitive health no matter who you are.

How is this relevant to rewiring?

Growth hormone interacts with other molecules such as IGF-1. Together they play critical roles in neuroplasticity.

Growth hormone enhances the release of dopamine involved in reward pathways and learning. Dopamine helps you remember what is pleasurable so your brain can encode that as a learned behaviour.

Growth hormone increases the excitability of neurons, making them more responsive to inputs, thus increasing the likelihood that new communication will stick.

Inflammation and immunity

Sleep is essential for maintaining a healthy immune system. When we sleep, the immune system eliminates the damaged cells and unwanted pathogens which contribute to illness. Have you ever noticed that when you're ill a good night's rest can help you feel better?

Studies have shown that individuals who sleep for six hours a night actually alter their genetic make-up to contribute to ill health, which can later down the line lead to neurodegeneration, diabetes and cardiovascular disease. In one study,[83] within one week people sleeping for six hours a night had altered approximately 3% of their genes negatively. Half those genes affected people's immune systems and increased inflammation in the body, and the other half increased the rate of tumour processing.

I'm telling you this because I want you to see how important sleep is and how when we don't prioritize it, it undermines our health drastically and keeps us in low-power mode. It also inhibits our ability to make new memories and really stay on top of our mental health. It affects all physiological processes down to our very genetic make-up.

Healthy adults spend around:

- Approx. **20–25%** of their time sleeping in **REM**; that's 100–120 minutes if you sleep for 7–8 hours
- Approx. **17–20%** of their night in **deep sleep**; that's 80–95 minutes if you sleep for 7–8 hours

Top tips on getting good sleep:

HOT SHOWERS HELP LOWER BODY TEMPERATURE

NO PHONES (SCROLLING/EMAILS/MESSAGES, ETC.) FIRST THING IN THE MORNING

AVOID CAFFEINE 8 HOURS BEFORE SLEEP

NO SCREENS 1 HOUR BEFORE BED

NO FOOD 2 HOURS BEFORE BED

BLACKOUT ROOM/EYE MASK

MAGNESIUM L-THREONATE

STRETCH BEFORE BED

MORNING SUNLIGHT

EVENING SUNLIGHT

MEDITATION

COOL ROOM

EXERCISE

REWIRING ESSENTIALS – SLEEP IS YOUR NUMBER-ONE OPTIMIZATION TOOL

- Sleep is arguably the most important step in creating new habits and making changes. It's where all our memories and newly acquired information becomes consolidated and encoded in the brain's memory centers.
- REM sleep is crucial for emotional processing and memory consolidation.
- REM sleep helps to remove the emotional load associated with intrusive thoughts.
- During REM, we release testosterone – a key regulator of cognitive function with neuroprotective properties.
- Deep sleep, the slowest-wave sleep, is the most restorative for our brains and bodies.
- It helps clear out toxins in the brain through the glymphatic system.
- There is evidence to suggest that improving deep sleep can contribute to insightful thinking and improve memory and learning.
- Growth hormone is released during deep sleep. It regulates metabolic functions such as insulin action and blood sugar levels, and it also helps with repairing cells in the brain that support neuron function.
- Growth hormone improves alertness, endurance and reduces irritability.

DOPAMINE DOESN'T WANT YOU TO BE HAPPY

DOPAMINE WANTS YOU TO HAVE MORE. IT'S ALWAYS ON TO THE NEXT THING. SO IF YOU'RE RELYING ON BEING HAPPY AFTER YOU GET WHAT YOU WANT, AFTER YOU ACHIEVE YOUR GOALS, AFTER YOU GET THE PROMOTION, YOU WILL ALWAYS BE CHASING HAPPINESS.

YOUR HAPPINESS IS NOW.

Dopamine – Your Happiness Is Now

You know that feeling? When you think, *As soon as I get to place B I will be happy*, or, *When I finally make x amount of money I will be happy*, but then you get the pay rise, you get the dream job, and it's not long before you realize you're still dissatisfied and start living for the next pay rise, because *that* is the one that's going to finally make you happy. Don't get me wrong, money helps, getting a pay rise certainly does make a difference, but there's a fine line between being happy because you're more financially stable and relying on the next goal achievement to be happy.

I asked my Instagram followers about whether they could relate to this feeling, and the most common themes were weight loss, work or academic accomplishments, and money. There were other examples too, like gaining Instagram followers, the number of likes on their social media posts and buying their first house. Some were professional athletes who never felt satisfied with any of their wins. The underlying theme is that whenever they got what they thought they wanted, they were underwhelmed and dissatisfied. The feeling of satisfaction lasted a short while before they realized they were still unhappy underneath, some feeling lost, lonely and depressed. I know this feeling all too well. I have always thought, *When I weigh x, I will love and be happy with my body.* But the arrival fallacy (a term coined by Dr Tal Ben-Shahar) is an illusion. It leads us into thinking that once we make it, once we get what we want and finally reach that goal, we'll finally be happy. When we strive for goals and progress towards them, the brain releases dopamine. But dopamine has been wrongly labelled up until recently. See, dopamine isn't the chemical of pleasure; it's the chemical that puts us in motivation drive *in pursuit of pleasure*. It is part of the reward pathway, and it does contribute to the experience of pleasure but it doesn't give you that feeling of satisfaction and reward. Instead it helps to reinforce pleasurable experiences by linking them to the desire of wanting to do them again, knowing they'll make you feel good.

Studies[84,85] show that when we achieve a goal, dopamine levels actually drop. This is because dopamine is responsible for anticipating a reward and driving you towards it; to maintain dopamine balance and avoid overstimulation, dopamine drops quickly after reaching your goal. The brain needs to reset the reward system to get prepared for future pursuits.

The pleasure actually comes from a cocktail of other neurotransmitters, such as serotonin, oxytocin and endorphins, which contribute to your feeling of success. As Dr Daniel Lieberman and Dr Michael Long beautifully explained in their book *The Molecule of More*, the cocktail of happiness is

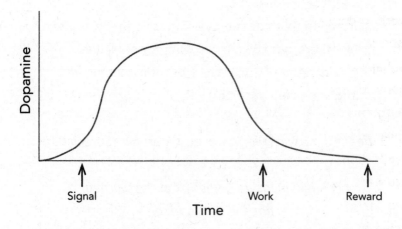

classified as the 'Here and Now' chemicals. They help you feel pleasure in the moment. But those neurotransmitters can easily be overrun by dopamine if we don't lean in to the moment and learn how to foster gratitude for what we currently have. This is why not everyone feels that sudden drop in happiness when they reach their goals. Some individuals feel a sense of accomplishment and satisfaction that lasts longer than it does in others.

If you're still unhappy regardless of your achievements, this could be grounded in many factors.

- Relying on goals to be happy
- Not enjoying the moment of accomplishment
- Going for the wrong goal
- Not addressing the root cause of unhappiness
- Masking unhappiness with goal setting
- Following a path because our peers pushed us to do so

- Living a life that's driven by a narrative that was given to us by others, not the one we want for ourselves
- Being depleted of dopamine from constant instant gratification (see page 281)

To beat the arrival fallacy, we must learn to appreciate that:

The real pleasure comes from the journey,
not the destination.

Happiness comes from the lessons you learn on the journey: the personal growth you achieve, the relationships you cultivate and the meaning of the journey. This aligns well with the growth mindset theory and the exercise we did on identifying our core values (page 119). When you stop worrying about the finishing line, you start living more joyfully.

The arrival fallacy says that we'll be happy when we reach the finishing line. But really there is no finishing line, and until you start believing that, you won't rest – because if you're waiting to get to the finishing line to be happy, you'll come to realize that there's another finishing line after that and another one after that . . .

The only race is against yourself.
Live easy in the moment. You won't get it again.

You might be asking yourself whether your motivation will

drop now that you have this information. Because if there isn't a real race, and no one is in this race with you, then what is the point? But the truth is, realizing this actually gives us a different perspective which can ignite a sense of determination and motivation to continuously improve, set higher standards and strive for personal growth and development. The focus shifts from comparing ourselves to others to becoming the best version of ourselves.

There is nothing wrong with using someone else as a bit of healthy competition. Perhaps your colleague is going for the same promotion, which motivates you to perform better at work. Yes, that's fair, so go for it. But your colleague isn't going to be there for every race. The main race. The ultimate race of YOUR life. Realizing that your primary competition is internal may prompt you to re-evaluate your goals and strategies to align better with your personal values, interests and abilities, leading to a renewed sense of purpose and motivation. More life.

ACTIVITY

What lessons have you learned on this journey to change?

_____

_____

_____

What is your meaning on this journey?

_____

_____

Boost the Positive

_____

What are you grateful for?

_____

_____

In what ways can you detach yourself from the achievement?

_____

_____

Have you gained transferable skills on this journey?

_____

_____

Move forward without fear of failure.
Move forward without regrets.
Because every experience is a lesson.

We've learned that our brain has a negativity bias. We're also prone to relying on arriving to be happy. This can be a recipe for disaster. The underlying theme here is living in the present and enjoying what lies directly in front of us. Taking time to appreciate the good in our lives instead of running on a hamster wheel and focusing on the negative. There is no such thing as real failure. Not really. Everything we do is a learning experience; every failure is an opportunity to rise from it. Neuroscience tells us that pleasure comes from the journey, not the destination, because there is no destination. Emphasizing personal growth and self-improvement will encourage introspection, self-awareness and a dedication to continuous learning and self-improvement which can lead you to a more purposeful life.

Instant gratification lowers your motivation to work towards goals

Dopamine is responsible for motivation and the pursuit of pleasure but it's also critical in seeking that pleasurable behaviour. It causes you to want, desire, seek and search for activities that will make you feel good. We call this reward-seeking behaviour, and when we find those activities, dopamine increases the likelihood of that behaviour being repeated in the future.

The pursuit of these pleasurable activities can drive us towards specific tasks and behaviours which lead to positive outcomes. For example, if you study to get good grades, chances are that you will study hard for future tests knowing that hard work pays off. Or you might engage in hobbies that bring you joy and satisfaction, especially if you start noticing a difference in your health. However, when such reward-seeking behaviour becomes excessive, it can lead to addiction, especially if the activities provide intense rewards, such as drug abuse and binge eating, which can lead to an unhealthy, cyclical pattern of reward seeking despite the adverse consequences.

Instant gratification refers to the immediate satisfaction

and pleasure that come from obtaining a reward without having to wait or put in much effort. It plays a significant role in reward-seeking behaviour and is the tendency to prioritize short-term satisfaction over more substantial rewards that require patience and effort. This can have a huge impact on our dopamine levels and the brain's reward system. Instant gratification can come from activities such as scrolling through social media, playing too many video games and by grabbing quick TV dinners and takeaways or eating highly processed and sugary foods. Another classic example is when a person engages in impulsive shopping despite their long-term financial goals of saving for a house. None of these things are bad when done in moderation; the issue is that these activities trigger a quick and intense release of dopamine that leads to the feeling of pleasure, but there is no effort involved.

The underlying rule for anyone, including those with ADHD, is that we need to exert effort in obtaining a reward. If we frequently rely on instant gratification to make ourselves feel better, we disrupt the brain's reward system and our dopamine levels. This is because our baseline dopamine changes and we need more to feel motivated to do something. Usually when your dopamine levels are at baseline, you'll feel complacent and not motivated to go after your goals. To feel motivated, your dopamine levels need to rise and this requires effort. The problem is that 'easy' doses of dopamine from social media, for example, eventually change your dopamine baseline, meaning you need more dopamine to feel motivated to do something.

Instant gratification lowers our motivation and desensitizes our brain's reward circuitry, leading us to need more dopamine to feel motivated to do anything. I know this feeling very well when I'm exercising. If I've been rampantly scrolling on social media prior to a workout, I slog through the whole ordeal, lacking the motivation to put one foot in front of the other. I usually give myself at least thirty minutes between social media and any self-led form of exercise (not in a class) to ensure that I get through the whole thing without being tempted to walk out every five minutes.

The constant and rapid surges in dopamine release from instant gratification mean that our brains become less responsive to dopamine and thus we find it harder to experience the same level of motivation from activities that require more effort. Activities such as exercise, pursuing long-term goals and sometimes even mundane things like leaving the house for grocery shopping, etc.

As your brain becomes accustomed to knowing it can get instant gratification from certain activities, it may start to favour these forms of reward over ones that require effort, especially if the effort needs to be sustained for a longer period of time. This is particularly important for those who have ADHD or low motivation in general. It's a catch-22 because scrolling and getting dopamine quickly lulls you into a false sense of motivation, but the dopamine hits are too volatile and therefore they send you crashing down quickly, lowering your motivation even more.

A RULE FOR DOPAMINE:

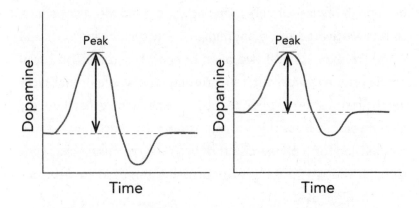

The larger the difference between baseline and peak (the delta), the more dopamine is released.

Although the peaks are the same, less dopamine is released because of the higher initial baseline.

It is suggested, for everyone, including those with ADHD, to put effort into attaining reward.

Activities such as exercise and cold-water exposure appear to increase dopamine in exponential ways without the immediate crash. Studies[86] have shown that cold-water exposure (submerging your body in cold water up to your neck) increases dopamine by about 250%, which takes up to three hours to come down. The water needs to be cold but tolerable and the individual needs to be mentally calm in order to benefit from it. Some people are able to go colder, and research has looked at temperatures below 14 degrees Celsius. Cold-water exposure isn't for everyone, and I understand that, but there are other ways to increase your dopamine.

Exercise

Running

Drawing

Cooking

Sunlight

Meditation

Focused work

Lifting weights

Experiencing new things

Basically whatever the activity is, it needs to have effort attached to it. Too many instant gratifications and we start disrupting our dopamine systems, putting us on a dangerous roller coaster of chasing that high. Next time you're scrolling on social media for hours on end, pay attention to how you feel. When I've spent too much time on social media, I start getting irritable and I actually get more addicted as time passes. I struggle to get off it and then I struggle to find the motivation to do anything else.

Our brains are wired to restore balance and therefore when your dopamine levels spike constantly, the brain may compensate for this, meaning we need bigger and more frequent dopamine hits to feel good. This diminishes our levels of motivation and pleasure for long-term goals. It is essential to maintain a healthy exposure to reward-seeking behaviours to avoid tolerance and addiction because these can ultimately lead to anxiety and depression. This can put us in a place where we feel unmotivated about life, which can spiral us into a

How to break the habit of scrolling

THINK OF YOUR BRAIN LIKE A MUSCLE. THE MORE YOU TRAIN IT, THE BETTER IT BECOMES

TIME-BLOCK A DEDICATED WORK TIME (START SMALL)

ALMOST 50% OF INTERRUPTIONS PEOPLE EXPERIENCE ARE SELF-INITIATED, BUT THAT CAN BE TRAINED

PHONE ON AEROPLANE MODE IN ANOTHER ROOM

MEDITATE TO IMPROVE FOCUS

LEARN TO IGNORE DISTRACTIONS AND REDIRECT YOUR THOUGHTS TO YOUR WORK – IT'S A SKILL

COLD EXPOSURE/EXERCISE HELPS RELEASE DOPAMINE

THERE ARE APP-BLOCKING APPS (YES THE IRONY, BUT THEY WORK)

trajectory of ruminating thoughts, which goes against the work we're trying to do around undoing the narratives that we've been repeating.

Harnessing boredom

Dr Anna Lembke, the bestselling author of *Dopamine Nation*, explained in her book that pain and pleasure are balanced in the same system. When we feel pleasure, shortly after we feel a pang of pain. This pain usually comes in the form of 'I want more'. You know that feeling of wanting to watch another episode, eat more chocolate, go back on holiday? That's dopamine wanting and seeking more. If the see-saw is strongly off balance so that we are swinging between the two, it can eventually lead to addiction. Have you ever left your phone at home and you had a dreadful feeling of wanting to reach for it when you knew it wasn't there? Phone addictions, predominantly in young adults, are very common. Studies[87] from 2022 estimate that the number is sitting at 61% in adolescents.

Provided someone isn't dealing with a substance addiction, boredom is what resets the see-saw. Sitting in the boredom of

not having your phone with you. Sitting in the boredom that allows you to access parts of your brain that take you on to a mind-wandering creative journey. We're always being influenced externally. When last did you daydream? By constantly disrupting our dopamine systems, needing more entertainment, needing something to do and something to keep us distracted, we're losing our ability to be bored. Being bored is a skill, and when you go through this phase, you may feel tempted to find other reward-seeking activities such as eating chocolate, spending money, online shopping. But it is wise to learn to become comfortable in the boredom of not engaging in anything dopaminergic and ensuring there is effort and delayed gratification when attaining pleasure and reward.

Not all tasks will be pleasurable, but some are non-negotiable.

Not all tasks become a habit, especially if they are effortful and unpleasant. But sometimes we have to do these things. For example, some people find exercise very enjoyable and can go alone, whereas others don't. My suggestion is to make it fun with friends. Stick to a routine and understand that this may be the case forever. We can't always rely on motivation to drive us. Some things are just non-negotiable.

REWIRING ESSENTIALS – DOPAMINE – YOUR HAPPINESS IS NOW

- Dopamine isn't the chemical of pleasure; it's the chemical that puts us in motivation drive *in pursuit of pleasure.*
- Studies show that when we achieve an anticipated goal, dopamine levels actually drop.
- The arrival fallacy deludes us into thinking that once we make it, once we get what we want and finally reach that goal, we'll finally be happy.
- Dopamine doesn't want you to be happy. Dopamine wants you to have more.
- If you're relying on being happy after you get what you want, after you achieve your goals, after you get the promotion, you will always be chasing happiness.
- The real pleasure comes from the journey, not the destination.
- Instant gratification lowers your motivation to work towards goals.
- When we prioritize immediate satisfaction and pleasure without having to wait or put in much effort, it can lead to an unhealthy, cyclical pattern of reward seeking.
- Instant gratification lowers our motivation and desensitizes our brain's reward circuitry, leading us to need more dopamine to feel motivated to do anything.
- A rule for dopamine: it is suggested, for everyone, to put effort into attaining reward.

Boost the Positive

- Prioritize exercise, goals, cooking, etc. versus excessively ordering takeout and avoiding doing focused work because it's easier to scroll social media.
- Harnessing boredom is the key to resetting reward circuitry.
- Pain and pleasure are anchored in the same system. After attaining reward, you'll feel some pain; this usually manifests itself in the form of 'I want another piece of chocolate'. Understanding that that's temporary and sitting with the discomfort will help reset the scale.

TRUST YOUR GUT:

PERSONAL ACCOUNTABILITY BUILDS
SELF-TRUST. SELF-TRUST BUILDS
CONFIDENCE.

Build Self-Trust and Confidence

You know that friend who always calls you and asks for advice, but is also asking ten other people as well and still doesn't listen to any of it? That was me. I also have a few friends like that. I was battling an internal argument, knowing the solution but not fully trusting in it. I think most people can relate to not fully knowing whether to trust their intuition. This feeling is grounded in many aspects. You may have had past negative experiences which influence your judgement in fear that something bad may happen again. Or you have doubts because you're an overthinker. You could also be fearful of the outcome or fearful of rejection. But the solution is grounded in building confidence through building self-trust.

Being confident is a powerful skill that can help us navigate our lives more positively, and we can learn it by implementing self-trust. Self-trust is having confidence in your abilities, building personal accountability and making sure you keep the promises you make to yourself. When you have confidence in your own abilities and control over your decisions and judgements, you create a strong foundation for personal growth, success and overall mental well-being.

Self-trust feeds in to all aspects of our lives, so we can learn to be more at ease with what's coming our way. When I got my first email from one of the biggest publishing houses in the world, I was elated. My sister, who was helping me with my emails, said to me, 'You may want to sit down for this one.' I was so happy my heart could have exploded. This opportunity had come to me because I was ready to put something out into the world without the panic of not knowing who I was and what I wanted. Had the opportunity come a year earlier, I don't know if I would have been ready for it. But given the transformation I had undergone in the years leading up to this, I was ready and I was calm. Happy and excited but calm.

I was preparing for my call with the editor and putting together a proposal for this book, letting my gut do the writing. It all came pouring out of me from beginning to end. I had never written something I was so proud of. It needed a lot of work, of course, but these were my original thoughts, my original ideas, all in one place. My confidence and self-trust, grounded in a growth mindset, allowed me to write something that didn't have any fear attached to it. I was not afraid of the outcome. I knew that if it didn't work out this time, it would work out at another time. Maybe other publishing houses would contact me and I would learn from this experience and do better next time. This level of self-trust allowed me to put out my best work and, for once, be really proud of it.

I built this level of trust and confidence by ensuring personal accountability. I took responsibility for my actions and decisions and their consequences. When I made promises to myself,

I kept them. When I said I was going to wake up at 6 a.m. and go for a run, I did it because I had promised myself I would. If I can't rely on myself, then how can I rely on anyone else? And how can I expect anyone else to rely on me?

How many times have you broken promises to yourself? Breaking the promises we make to ourselves undermines the relationship we have with ourselves and, in turn, our confidence in the world.

STEPS TO BUILDING SELF-TRUST

Honour your commitments. When you make promises to yourself and stick to them, you reinforce self-trust. This can be completing a task on your to-do list that you said you were going to do or going to the gym when you said you were going to go.

Setting realistic goals. This means that the commitments you make with yourself need to be realistic and achievable. You also need to stay true to yourself and set goals and commitments that align with your values. For example, you may have goals that you haven't set for yourself; perhaps they are goals that you think you *should* be doing. Or perhaps you're setting goals that are too hard to achieve, which will inevitably end in failure. Be realistic and set goals that truly align with you.

Celebrate your accomplishments. Even the minor ones. Our brains tend to dwell on the negative, so make sure you celebrate your wins so that you can start shifting your narrative to see the positives in yourself. This will strengthen your self-trust so that you're able to recognize your capacity for success.

Growth mindset. Remember, every failure is a learning opportunity. You either win or you learn. So instead of dwelling on your failures, focus on how you can improve.

Listen to yourself. Pay attention to your thoughts and feelings. Trust your gut feeling. Listening to your intuition and acting on it is going to reinforce self-trust so that you're able to navigate life knowing you can advocate for yourself. Sometimes this means carving out time for yourself to meditate/stretch/walk or any activity that's going to disengage you from the external world. Remember the importance of carving out mental breaks that allow us to access parts of our brain that give us clarity (see page 71).

Self-trust and trusting your intuition are closely linked concepts that play significant roles in decision-making, personal growth and overall well-being. Trusting your intuition requires a foundation of self-trust. If you're able to trust yourself, you're able to use past experiences your brain has stored as signals. When you lean in to your intuition, you start seeing that some of the things you're feeling you have experienced before and you can make better decisions. For example, you're in a new relationship and this person is starting to show you red flags. Your intuition is telling you that something is off because you have experienced something similar before. Your brain knows that deep down this is wrong, even though your conscious brain is trying to rationalize the reasons to overlook them.

Your gut feeling is the result of your subconscious mind making decisions without your conscious awareness. Studies[88] show that we can still make decisions even when we're not consciously thinking about it. Your subconscious brain is a network of intertwined neurons that communicate with each other constantly. Those thoughts do not arise all the time because your conscious brain has a limited capacity for what it can hold in mind. Imagine your subconscious constantly reminding you of every memory you've ever had; you wouldn't have space to think about anything else. The memories are still there, though, and sometimes they arise as a gut feeling, something that you haven't quite made sense of yet – until that lightbulb moment when it's painfully obvious that you knew the answer all along.

When you trust yourself, you're more likely to trust the signals your intuition sends you. And it's a reciprocal relationship because when you rely on your intuition and see positive results, it reinforces your self-trust by showing that you have an innate voice speaking to you. Self-trust and trusting your intuition can work together to drive you towards decisions that align with who you really are and who you want to be. And the more you lean in to it, the more your intuition will become clearer and more reliable.

> *'Most of the problems in my life have come from lettin'*
> *my head get too far ahead from my heart'*
> – Matthew McConaughey

Trusting your intuition is single-handedly the best thing you can ever do for yourself. And I appreciate that even if your intuition is strong, you might wonder whether you're about to make a huge mistake. But learning to trust your innate intuition means you begin to live a life where your head and your heart work as one. A life where an internal compass guides you through the labyrinth of existence. Through the hard times and the easy times. Trust what your heart is saying and let your head complement this wisdom. In this beautiful collaboration you'll find the courage to follow your dreams, the resilience to overcome challenges and the inspiration to create yourself. Knowing this means you can go forth with less fear, because you'll know that when all else fails, when everybody disagrees with you, you'll have your own back. And is that internal alignment not the best thing that we could ask for?

ACTIVITY

Reflecting on recent events, perhaps relationships, what red flags have you chosen to ignore when you should have trusted your intuition?

_____

_____

_____

What did you learn from that experience? Can you see that your gut feeling is a form of self-trust?

_____

_____

Going through this process of self-trust and keeping promises to yourself might be uncomfortable at first. There may be points where you'll want to revert to automatic. And maybe you will for a while because you'll have good days and you'll have bad days. But don't forget that the brain doesn't unlearn everything in a day, just like it doesn't learn it all in a day either. The brain is plastic, not elastic, and change isn't linear.

**IF YOU BREAK YOUR PROMISES FOR A FEW DAYS,
DON'T WORRY.**

DON'T BEAT YOURSELF UP ABOUT IT.

ASSESS WHY YOU'RE DOING IT.

REMEMBER THAT THE BRAIN LIKES TO REVERT TO
AUTOMATIC. ACKNOWLEDGING THIS HELPS US
UNDERSTAND WHY WE PUSH THROUGH
RESISTANCE WHEN MAKING CHANGES FOR
OURSELVES.

GET BACK ON TRACK WITHOUT DWELLING ON
THE NEGATIVE.

REMIND YOURSELF WHY YOU WANT TO BUILD
SELF-TRUST.

If you've been putting in the work and you start slipping up, remember that you won't forget it all overnight. It won't suddenly disappear because the groundwork you've laid will still be there. Be patient with yourself and follow the steps above when this happens.

Remember that mental heuristics are shortcuts that your brain wants to take in order to save energy for more cognitively demanding tasks, and it doesn't care all that much about change. Yes, the brain is designed to change, but it's mostly designed to keep you alive. So change can sometimes feel sticky. Let's say you want to stop grabbing your phone first thing in the morning. You need to wake up every day and remind yourself that you want to do that. At first this is going to feel uncomfortable because your brain is going to want to revert to automatic, and you may even wake up and forget you even wanted to try, but don't fret – habits are not built overnight.

There are a few things you can do.

ONE

Practise visualizing yourself waking up without grabbing your phone the night before. Think about all the steps involved in you waking up and not scrolling on social media. What will you do instead? How will you feel when you know you're going to want to do it anyway? Because of mental heuristics your brain is practising what it knows best. Remind yourself of dopamine's desire for short-term rewards and how that may impact you in the morning. Bearing all these things in mind means that we can start having more compassion for ourselves.

Many people believe that they have no willpower, or that it's in their personality type to be a certain way. But when we empower ourselves with knowledge, we take back control and see that we can indeed break these patterns. Remember what we learned about ditching the negative and the steps involved in severing a trigger and the response? Currently your brain is operating on a communication bond that's very strong: trigger (waking up and wanting the phone) and response (automatically reaching for it). Over time this bond will get weaker if you leave enough space between the trigger and the response.

TWO

The second thing you can do is to change your habit cues (see Chapter 7). It would be much easier for you to break a morning phone-grabbing routine if you moved to a new country because it would give your brain the opportunity to start fresh. So changing things up is an effective way to make sure you break a pattern like this. Perhaps buying an alarm clock and putting your phone in a different room on aeroplane mode. Perhaps changing your alarm or putting your phone in another part of your room. Choose the more extreme option (phone in a different room) if you think you'll break the promise to yourself about grabbing it to wake up.

Sometimes I wake up in the morning with the urge to grab my phone, but I fight against that automatic behaviour because I know that eventually waking up without that urge will be the new habit. Creating new habits is the new black. But you have to push through the discomfort until the wiring sequence of neurons that fire in arrangement one after the other no longer communicates in that way, until a new sequence has been implemented and you wake up not caring that your phone is burning with notifications. Beating yourself up gives you a false sense of control – you can't change the past – but by beating yourself up you can feel better about knowing better. Remember, the brain wants to take the path of least resistance. We just have to be patient as we carve out the new routes.

Having strong self-trust means that you will possess more confidence, which means that you can start to take on more challenges while embracing new opportunities without fear. This allows you to take ownership of your actions and decisions, which cultivates a sense of control over your life. These qualities coupled with a growth mindset can help you to pursue your dreams and live a life filled with bravery and determination.

For me, building self-trust means that I have some non-negotiables. It means I do things as a routine, despite the fact that I am lacking motivation. Motivation is temporary, and it was when I learned that it disappears for everyone that I really started to value the pursuit of pushing through the grit until something becomes ingrained. For example, most days, if I really and truly think about it, I don't want to leave the house or exercise. It's so much easier to get lost in my work and call it 'too busy'. But when I get up at 6 a.m. and move my body in some way or another, I always feel so much better. That's what works for me, as I am a morning person. You might not be and that's fine, but for me it means that the rest of my day clicks into place and follows a much smoother trajectory than if I sleep in and snooze. Don't get me wrong, I do believe that lazy days are important, but I have learned over many years what my limits are. For some it might be waking up at 8 a.m. and walking instead of running, and that's absolutely fine. The point is that you find what your non-negotiables are and you stick to them because that way you'll reinforce self-trust. Start with achievable goals and then add on from there as needed.

Motivation will only get you so far, and then you need to rely on discipline and routine. Like clockwork, the cogs need to keep ticking in repetition to wire the brain the way you want it to. I promise you that your motivation will dissipate, so don't get used to it, because that's why people usually fail at making any change.

What are your non-negotiables?

Mine are:

- Exercise every day Monday to Friday, even if it's just a stretch or a walk with the dogs. It doesn't always have to be at 6 a.m. but, unless I'm unwell, at some point in the day I carve out time to move my body. The intensity can vary.
- Switch phone off at 9 p.m.
- Get eight hours of sleep.
- Meditate/visualize/self-hypnotize every day even if it's just for one minute.
- Limit social media use as much as possible (to refocus my mind and avoid self-interruption).
- Don't check my phone first thing in the morning.
- Swim in the sea at least twice a week.

What are yours?

_____

_____

_____

How often do you want to achieve these per week?

_____

_____

_____

Can you see the benefit in creating non-negotiables?

_____

_____

_____

Do you believe that this can lead you to trusting yourself better?

_____

_____

_____

REWIRING ESSENTIALS – BUILD SELF-TRUST AND CONFIDENCE

- Self-trust is having confidence in your abilities, building personal accountability and making sure you keep the promises you make to yourself.
- Having self-trust boosts confidence.
- Building self-trust starts with honouring your commitments and keeping promises to yourself.
- Celebrating your accomplishments will strengthen your self-trust so that you're able to recognize your capacity for success.
- Pay attention to your thoughts and feelings.
- Trust your gut feeling. Listening to your intuition and acting on it is going to reinforce self-trust by knowing you can advocate for yourself.
- Your gut feeling is the result of your subconscious mind making decisions without your conscious awareness.
- Studies show that we can still make decisions even when we're not consciously thinking about it.
- When you trust yourself, you're more likely to trust the signals your intuition sends you from your subconscious.
- Motivation is temporary, so having non-negotiables is important in ensuring we stick to the things that are good for us through discipline.
- Find what your non-negotiables are and stick to them because they will reinforce self-trust.

Outro

Power

I've answered so many questions since holding that first brain in my hands. There are some that I'll never be able to answer, like whether that brain I held was happy, but what I do know is that we are given a unique set of gifts in this lifetime. The traits and qualities that are unique to you have been dictated by the idiosyncratic genes coded into your DNA. The brain maps of your internal world that no one will ever be able to replicate, no matter how much we try to blend in with the others. Use these gifts. Exhume these traits. Share them with world. Use them to tell a better story about yourself. Declaring your métiers to the world isn't a selfish act that we should shy away from. Tell your story. Tell a better story . . . It's your God-given right.*

I'll never know whether the brain I held knew how to break free from ruminating thoughts, but I hope that they were

---

* I mean that despite what society wants you to be and the way it conspires against us, despite the narratives you've been told and despite the trajectory you were set on . . . there are certain qualities within us that are the birthright of every living person on this Earth.

happy. And I do know the brain can change. We can change. We can be whoever we want to be and we can create ourselves from scratch. Reprogramme ourselves to live a fulfilling life free from the chains of ruminating thoughts and the bad habits that hinder us.

Our DNA tells us that we are given a unique set of gifts in this lifetime, and it's our duty to uncover them and share them with the world. Only life, expectations, conditioning and societal pressure tell us we aren't capable, we aren't good enough or that we have in some way messed it up or left it too late. Find those qualities and amplify them into greatness. And then help the people around you. Greatness is finding your own qualities, and then influencing those around you to find theirs. Greatness is learning who you are and then passing it on to the next person so that they can pass it on to the next one, asking for nothing in return. Infiltrating other people with your energy through guidance and actions. Remember, love invents us.

So let me ask the question again: is it by mistake or design? It's both.

Commitment

By virtue of the associations between the synapses that communicate with one another in your brain, you can create your own mind and be whoever you want to be.

By virtue of the associations between the neurons in your brain, you're capable of anything.

Create yourself.

Commit to yourself.

Shake hands on it with yourself.

Set yourself free of this feeling that you're stuck. You've spent a long time running, in fear; but fear is not your enemy. Fear is your reason to do it. And when you accept the fear, you learn that it wasn't all that scary in the first place. The world will commend your courage. Courage is not the absence of fear but rather the triumph over it. It is acknowledging fear yet choosing to move forward with purpose and conviction.

Hurl yourself into the darkness with courage and come to learn that *you* are the light that allows you to see. A light that is joyful and helps you live without negative self-beliefs that whisper you're not good enough.

Hurl yourself into the unknown and come to learn that every bit of knowledge you need is at *your* fingertips, that the obstacle you were afraid to climb is, in fact, the first step that took you over the line. Give yourself and your brain more credit, because the only thing that's stopping you right now is you.

Everything you've been through has prepared you for this. And when things start falling apart again, because they will, remember that you're equipped to deal with the adversities that life throws your way.

Face your fears, push through the unknown ... you never know what's on the other side. Perhaps it's failure, perhaps it's not. Perhaps you'll believe me when I tell you that there is no such thing as failure. Not really anyway. So fail away – it's all a learning experience. The race is against yourself.

Don't succumb to comparison, gossip and doom-scrolling.

Live your life.

Prove them wrong.

You always hold the power to change your destiny, to alter the preprogrammed course of your life to something that's dictated by you and you alone.

Equipped with the knowledge of neuroplasticity . . . go forth.

Forgiveness

There is one more thing you need to do.

The last step on your journey of change, arguably the hardest step . . . is forgiveness.

Redeeming ourselves of the self-limiting beliefs instilled in us during our formative years requires forgiveness towards those who unknowingly or unintentionally programmed us. It begins by empathizing with their intentions and circumstances, understanding that they may have been operating within the confines of their own knowledge and experiences. By shifting the focus from blame to self-responsibility, we recognize the fallibility of the individuals who influenced us, who were driven by their own fears and limitations. By acknowledging the power of their influence and our ability to reprogramme ourselves, we release the grip of resentment, allowing room for personal growth and healing. Forgiveness grants us the peace and freedom needed to pursue our potential unburdened by the weight of past beliefs, empowering us to shape a brighter future.

Without forgiveness we run the risk of becoming the people who hurt us.

May this book be the light you needed at the end of the tunnel. Everything you need to make a change is within this book, and, having read it, it is now within you.

Don't fight against time – good things take time. This book has been designed so that you can come back to it at any point, at any given time. You're the greatest project you'll ever work on. Don't hold back.

Equipped with the knowledge to change your life . . . close this book and look up to the sky . . . There's a world out there and you can be whoever you want in it with your new narrative.

Create yourself.

Notes

A note on autism

One of the leading theories about individuals with autism spectrum disorder (ASD) is that they have atypical connectivity in the brain. For some that connectivity might be exaggerated, meaning that there is an excessive or increased level of connectivity between neurons or regions of the brain – hyperconnectivity – and for some there is a reduction – hypoconnectivity. These theories are being explored as potential factors that contribute to the characteristics of autism.

The hyperconnectivity hypothesis suggests that there may be an overgrowth of neural connections in certain areas of the brain, leading to enhanced processing of specific types of information but potentially impairing the integration of others. This could contribute to the uneven cognitive and sensory profiles often seen in individuals with autism. However, this means that while people with ASD may experience some challenges, they also have a unique set of strengths. Individuals with ASD are capable of neuroplasticity, which means that individuals with ASD are capable of creating new habits and changing behaviours should they wish to.

A note on OCD

One of the key areas of the brain for individuals with OCD is the caudate nucleus. It usually facilitates the transition of focus to the next task; however, individuals with OCD often experience impaired caudate function, a significant aspect of the disorder. This deficiency hampers the ability to shift attention smoothly. Addressing this issue involves actively engaging in the cultivation of alternative thought patterns, challenging the diminished functionality of the caudate. By learning and reinforcing new cognitive approaches, individuals aim to counteract the challenges that one usually experiences, fostering the potential for improved mental health and coping mechanisms.

Therapeutic interventions, such as CBT, capitalize on neuroplasticity by helping individuals reshape thought patterns and behaviours. Through targeted exposure and response prevention, the brain learns alternative, healthier responses to obsessive thoughts. This rewiring process contributes to symptom reduction and improved emotional well-being.

So can this book help someone with OCD? Absolutely.

A note on ADHD

ADHD can pose some considerable challenges to individuals striving to change their habits and behaviours. However, there is a major misunderstanding in individuals with ADHD whereby people think that ADHDers have short attention spans or an inability to concentrate. A better way to think about it is that ADHDers have an asynchrony in the attentional networks of the brain whereby they tend to prioritize the tasks they find most psychologically rewarding and block out the rest of the world around them. This is because individuals with ADHD possess the ability to hyperfocus on the things that they really care about. This hyperfocus means that individuals often become so engrossed in their tasks that they fixate on them and struggle to move on to something else, especially when it's unrewarding.

This hyperfocus can be a blessing for achieving goals and dreams that suit their endeavours, but it can be a curse because it means that ADHDers struggle to stick to boring but necessary tasks that are a requirement for everyday life. This can lead to impulsivity, poor time management and prioritization of tasks. Additionally, motivation and reward sensitivity also play a role, with individuals with ADHD often grappling with sustaining motivation and being more responsive to immediate rewards. This heightened sensitivity may make it challenging to find the necessary motivation for long-term benefits, making it harder to remain committed to habit-building efforts. Despite these hurdles, individuals with ADHD can successfully navigate habit change by implementing tailored strategies, including breaking tasks into manageable steps, setting clear goals, using reminders and organizational tools, and seeking support from appropriate mental health professionals. The ADHD brain is capable of neuroplasticity, which means that the frontal cortex,

Notes

the area of the brain responsible for executive function can be strengthened. It's important for ADHDers to use positive reinforcement when achieving small goals and ensuring that they celebrate the small wins in life when it comes to mundane and boring tasks, and when achieving the small habits they are trying to change. These small celebrations can aid in dopaminergic reward activity which can help solidify neuronal changes in key brain areas.

References

1. Koch, G., & Spampinato, D. (2022). Alzheimer disease and neuroplasticity. *Handbook of Clinical Neurology, 184*, 473–479. https://doi.org/10.1016/B978-0-12-819410-2.00027-8

2. Casaletto, K., Ramos-Miguel, A., VandeBunte, A., Memel, M., Buchman, A., Bennett, D., & Honer, W. (2022). Late-life physical activity relates to brain tissue synaptic integrity markers in older adults. *Alzheimer's & Dementia: The Journal of the Alzheimer's Association, 18*(11), 2023–2035. https://doi.org/10.1002/alz.12530

3. Recanzone, G. H., Merzenich, M. M., & Jenkins, W. M. (1992). Frequency discrimination training engaging a restricted skin surface results in an emergence of a cutaneous response zone in cortical area 3a. *Journal of Neurophysiology, 67*(5), 1057–1070. https://doi.org/10.1152/jn.1992.67.5.1057

4. Jean Liedoff: 'The Continuum Concept: In Search of Happiness Lost'

5. Lally, P., van Jaarsveld, C. H. M., Potts, H. W. W., & Wardle, J. (2010). How are habits formed: Modelling habit formation in the real world. *European Journal of Social Psychology, 40*(6), 998–1009. https://doi.org/10.1002/ejsp.674

6. Crum, A. J., Salovey, P., & Achor, S. (2013). Rethinking stress: The role of mindsets in determining the stress response.

Journal of Personality and Social Psychology, 104(4), 716–733. https://doi.org/10.1037/a0031201

7. Jamieson, J. P., Nock, M. K., & Mendes, W. B. (2012). Mind over matter: Reappraising arousal improves cardiovascular and cognitive responses to stress. *Journal of Experimental Psychology. General, 141*(3), 417–422. https://doi.org/10.1037/a0025719

8. Wiehler, A., Branzoli, F., Adanyeguh, I., Mochel, F., & Pessiglione, M. (2022). A neuro-metabolic account of why daylong cognitive work alters the control of economic decisions. *Current Biology : CB, 32.* https://doi.org/10.1016/j.cub.2022.07.010

9. Payne, P., Levine, P. A., & Crane-Godreau, M. A. (2015). Somatic experiencing: Using interoception and proprioception as core elements of trauma therapy. *Frontiers in Psychology, 6* https://www.frontiersin.org/journals/psychology/articles/10.3389/fpsyg.2015.00093

10. Balban, M. Y., Neri, E., Kogon, M. M., Weed, L., Nouriani, B., Jo, B., Holl, G., Zeitzer, J. M., Spiegel, D., & Huberman, A. D. (2023). Brief structured respiration practices enhance mood and reduce physiological arousal. *Cell Reports Medicine, 4*(1), 100895. https://doi.org/10.1016/j.xcrm.2022.100895

11. Stress-Proof: The Scientific Solution to Protect Your Brain and Body – and be More Resilient Every Day – Mithu Storoni

12. Cunha, C., Brambilla, R., & Thomas, K. (2010). A simple role for BDNF in learning and memory? *Frontiers in Molecular Neuroscience, 3.* https://www.frontiersin.org/articles/10.3389/neuro.02.001.2010

13. Szabadi, E. (2018). Functional Organization of the Sympathetic Pathways Controlling the Pupil: Light-Inhibited and Light-Stimulated Pathways. *Frontiers in Neurology, 9,* 1069. https://doi.org/10.3389/fneur.2018.01069

14. Alipour, A., Arefnasab, Z., & Babamahmoodi, A. (2011). Emotional Intelligence and Prefrontal Cortex: A Comparative Study Based on Wisconsin Card Sorting Test (WCST). *Iranian Journal of Psychiatry and Behavioral Sciences, 5*(2), 114–119.

15. Satpute, A. B., & Lindquist, K. A. (2021). At the Neural Intersection Between Language and Emotion. *Affective Science, 2*(2), 207–220. https://doi.org/10.1007/s42761-021-00032-2

16. Lieberman, M. D., Eisenberger, N. I., Crockett, M. J., Tom, S. M., Pfeifer, J. H., & Way, B. M. (2007). Putting feelings into words: Affect labeling disrupts amygdala activity in response to affective stimuli. *Psychological Science, 18*(5), 421–428. https://doi.org/10.1111/j.1467-9280.2007.01916.x

17. Unwinding Anxiety: New Science Shows How to Break the Cycles of Worry and Fear to Heal Your Mind – Judson Brewer

18. de Voogd, L. D., Kanen, J. W., Neville, D. A., Roelofs, K., Fernández, G., & Hermans, E. J. (2018). Eye-Movement Intervention Enhances Extinction via Amygdala Deactivation. *The Journal of Neuroscience: The Official Journal of the Society for Neuroscience, 38*(40), 8694–8706. https://doi.org/10.1523/JNEUROSCI.0703-18.2018

19. Lederbogen, F., Kirsch, P., Haddad, L., Streit, F., Tost, H., Schuch, P., Wüst, S., Pruessner, J. C., Rietschel, M., Deuschle, M., & Meyer-Lindenberg, A. (2011). City living and urban

upbringing affect neural social stress processing in humans. *Nature, 474*(7352), 498–501. https://doi.org/10.1038/nature10190

20. Sudimac, S., Sale, V., & Kühn, S. (2022). How nature nurtures: Amygdala activity decreases as the result of a one-hour walk in nature. *Molecular Psychiatry, 27*(11), 4446–4452. https://doi.org/10.1038/s41380-022-01720-6

21. Forys, W. J., & Tokuhama-Espinosa, T. (2022). The Athlete's Paradox: Adaptable Depression. *Sports, 10*(7), 105. https://doi.org/10.3390/sports10070105

22. Blain, B., Hollard, G., & Pessiglione, M. (2016). Neural mechanisms underlying the impact of daylong cognitive work on economic decisions. *Proceedings of the National Academy of Sciences, 113*(25), 6967–6972. https://doi.org/10.1073/pnas.1520527113

23. Albulescu, P., Macsinga, I., Rusu, A., Sulea, C., Bodnaru, A., & Tulbure, B. T. (2022). 'Give me a break!' A systematic review and meta-analysis on the efficacy of micro-breaks for increasing well-being and performance. *PLOS ONE, 17*(8), e0272460. https://doi.org/10.1371/journal.pone.0272460

24. Wiens, S., Sand, A., Norberg, J., & Andersson, P. (2011). Emotional event-related potentials are reduced if negative pictures presented at fixation are unattended. *Neuroscience Letters, 495*(3), 178–182. https://doi.org/10.1016/j.neulet.2011.03.042

25. Kensinger, E., & Schacter, D. (2006). Processing emotional pictures and words: Effects of valence and arousal. *Cognitive,*

Affective & Behavioral Neuroscience, 6, 110–126. https://doi. org/10.3758/CABN.6.2.110

26. Shouval, H., Wang, S., & Wittenberg, G. (2010). Spike Timing Dependent Plasticity: A Consequence of More Fundamental Learning Rules. *Frontiers in Computational Neuroscience, 4*. https://www.frontiersin.org/articles/10.3389/ fncom.2010.00019

27. Moutsiana, C., Garrett, N., Clarke, R. C., Lotto, R. B., Blakemore, S.-J., & Sharot, T. (2013). Human development of the ability to learn from bad news. *Proceedings of the National Academy of Sciences, 110*(41), 16396–16401. https:// doi.org/10.1073/pnas.1305631110

28. Karlsson, N., Loewenstein, G., & Seppi, D. (2009). The ostrich effect: Selective attention to information. *Journal of Risk and Uncertainty, 38*(2), 95–115. https://doi.org/10.1007/ s11166-009-9060-6

29. Sharot, T., Rollwage, M., Sunstein, C. R., & Fleming, S. M. (2023). Why and When Beliefs Change. *Perspectives on Psychological Science, 18*(1), 142–151. https://doi. org/10.1177/17456916221082967

30. Chou, T., Deckersbach, T., Dougherty, D. D., & Hooley, J. M. (2023). The default mode network and rumination in individuals at risk for depression. *Social Cognitive and Affective Neuroscience, 18*(1), nsad032. https://doi. org/10.1093/scan/nsad032

31. Garrison, K. A., Zeffiro, T. A., Scheinost, D., Constable, R. T., & Brewer, J. A. (2015). Meditation leads to reduced default mode network activity beyond an active task. *Cognitive, Affective &*

Behavioral Neuroscience, 15(3), 712–720. https://doi.
org/10.3758/s13415-015-0358-3

32. Sridharan, D., Levitin, D. J., & Menon, V. (2008). A critical role
for the right fronto-insular cortex in switching between
central-executive and default-mode networks. *Proceedings
of the National Academy of Sciences of the United States of
America, 105*(34), 12569. https://doi.org/10.1073/
pnas.0800005105

33. Pascual-Leone, A., Nguyet, D., Cohen, L. G., Brasil-Neto, J. P.,
Cammarota, A., & Hallett, M. (1995). Modulation of muscle
responses evoked by transcranial magnetic stimulation during
the acquisition of new fine motor skills. *Journal of
Neurophysiology, 74*(3), 1037–1045. https://doi.org/10.1152/
jn.1995.74.3.1037

34. Crum, A. J., Corbin, W. R., Brownell, K. D., & Salovey, P. (2011).
Mind over milkshakes: Mindsets, not just nutrients, determine
ghrelin response. *Health Psychology: Official Journal of the
Division of Health Psychology, American Psychological
Association, 30*(4), 424–429; discussion 430-431. https://doi.
org/10.1037/a0023467

35. Markmann, M., Lenz, M., Höffken, O., Steponavičiūtė, A.,
Brüne, M., Tegenthoff, M., Dinse, H. R., & Newen, A. (2023).
Hypnotic suggestions cognitively penetrate tactile perception
through top-down modulation of semantic contents.
Scientific Reports, 13(1), 6578. https://doi.org/10.1038/
s41598-023-33108-z

36. Kleck, R., & Strenta, A. (1985). Gender and Responses to
Disfigurement in Self and Others. *Journal of Social and*

Clinical Psychology, *3*, 257–267. https://doi.org/10.1521/
jscp.1985.3.3.257

37. Boothby, E. J., Cooney, G., Sandstrom, G. M., & Clark, M. S.
(2018). The Liking Gap in Conversations: Do People Like Us
More Than We Think? *Psychological Science*, *29*(11), 1742–1756.
https://doi.org/10.1177/0956797618783714

38. Eisenberger, N. I., Lieberman, M. D., & Williams, K. D. (2003).
Does rejection hurt? An FMRI study of social exclusion.
Science (New York, N.Y.), *302*(5643), 290–292. https://doi.
org/10.1126/science.1089134

39. Sturgeon, J. A., & Zautra, A. J. (2016). Social pain and physical
pain: Shared paths to resilience. *Pain Management*, *6*(1),
63–74. https://doi.org/10.2217/pmt.15.56

40. De Ridder, D., Adhia, D., & Vanneste, S. (2021). The anatomy
of pain and suffering in the brain and its clinical implications.
Neuroscience & Biobehavioral Reviews, *130*, 125–146. https://
doi.org/10.1016/j.neubiorev.2021.08.013

41. Verhallen, A. M., Renken, R. J., Marsman, J.-B. C., & ter Horst,
G. J. (2021). Working Memory Alterations After a Romantic
Relationship Breakup. *Frontiers in Behavioral Neuroscience*,
15, 657264. https://doi.org/10.3389/fnbeh.2021.657264

42. Orr, C., & Hester, R. (2012). Error-related anterior cingulate
cortex activity and the prediction of conscious error
awareness. *Frontiers in Human Neuroscience*, *6*, 177. https://
doi.org/10.3389/fnhum.2012.00177

43. Song, H., Zou, Z., Kou, J., Liu, Y., Yang, L., Zilverstand, A.,
d'Oleire Uquillas, F., & Zhang, X. (2015). Love-related changes
in the brain: A resting-state functional magnetic resonance

imaging study. *Frontiers in Human Neuroscience, 9,* 71. https://doi.org/10.3389/fnhum.2015.00071

44. Grebe, N. M., Kristoffersen, A. A., Grøntvedt, T. V., Emery Thompson, M., Kennair, L. E. O., & Gangestad, S. W. (2017). Oxytocin and vulnerable romantic relationships. *Hormones and Behavior, 90,* 64–74. https://doi.org/10.1016/j.yhbeh.2017.02.009

45. Weiskittle, R. E., & Gramling, S. E. (2018). The therapeutic effectiveness of using visual art modalities with the bereaved: A systematic review. *Psychology Research and Behavior Management, 11,* 9–24. https://doi.org/10.2147/PRBM.S131993

46. Tseng, J., & Poppenk, J. (2020). Brain meta-state transitions demarcate thoughts across task contexts exposing the mental noise of trait neuroticism. *Nature Communications, 11,* 3480. https://doi.org/10.1038/s41467-020-17255-9

47. Primack, B. A., Shensa, A., Sidani, J. E., Whaite, E. O., Lin, L. yi, Rosen, D., Colditz, J. B., Radovic, A., & Miller, E. (2017). Social Media Use and Perceived Social Isolation Among Young Adults in the U.S. *American Journal of Preventive Medicine, 53*(1), 1–8. https://doi.org/10.1016/j.amepre.2017.01.010

48. Riehm, K. E., Feder, K. A., Tormohlen, K. N., Crum, R. M., Young, A. S., Green, K. M., Pacek, L. R., La Flair, L. N., & Mojtabai, R. (2019). Associations Between Time Spent Using Social Media and Internalizing and Externalizing Problems Among US Youth. *JAMA Psychiatry, 76*(12), 1266–1273. https://doi.org/10.1001/jamapsychiatry.2019.2325

49. Lustenberger, C., Boyle, M. R., Foulser, A. A., Mellin, J. M., & Fröhlich, F. (2015). Functional role of frontal alpha oscillations

in creativity. *Cortex*, *67*, 74–82. https://doi.org/10.1016/j.cortex.2015.03.012

50. Westbrook, A., Ghosh, A., van den Bosch, R., Määttä, J. I., Hofmans, L., & Cools, R. (2021). Striatal dopamine synthesis capacity reflects smartphone social activity. *iScience*, *24*(5), 102497. https://doi.org/10.1016/j.isci.2021.102497

51. Ruffino, C., Truong, C., Dupont, W., Bouguila, F., Michel, C., Lebon, F., & Papaxanthis, C. (2021). Acquisition and consolidation processes following motor imagery practice. *Scientific Reports*, *11*(1), 2295. https://doi.org/10.1038/s41598-021-81994-y

52. Monany, D. R., Lebon, F., Dupont, W., & Papaxanthis, C. (2022). Mental practice modulates functional connectivity between the cerebellum and the primary motor cortex. *iScience*, *25*(6). https://doi.org/10.1016/j.isci.2022.104397

53. Son, S. M., Yun, S. H., & Kwon, J. W. (2022). Motor Imagery Combined With Physical Training Improves Response Inhibition in the Stop Signal Task. *Frontiers in Psychology*, *13*. https://www.frontiersin.org/journals/psychology/articles/10.3389/fpsyg.2022.905579

54. Iding, A. F. J., Kohli, S., Dunjic Manevski, S., Sayar, Z., Al Moosawi, M., & Armstrong, P. C. (2023). Coping with setbacks as early career professionals: Transforming negatives into positives. *Journal of Thrombosis and Haemostasis*, *21*(7), 1689–1691. https://doi.org/10.1016/j.jtha.2023.04.019

55. Hariri, A. R., Brown, S. M., Williamson, D. E., Flory, J. D., de Wit, H., & Manuck, S. B. (2006). Preference for Immediate over Delayed Rewards Is Associated with Magnitude of Ventral

Striatal Activity. *The Journal of Neuroscience, 26*(51), 13213–13217. https://doi.org/10.1523/JNEUROSCI. 3446-06.2006

56. Kox, M., van Eijk, L. T., Zwaag, J., van den Wildenberg, J., Sweep, F. C. G. J., van der Hoeven, J. G., & Pickkers, P. (2014). Voluntary activation of the sympathetic nervous system and attenuation of the innate immune response in humans. *Proceedings of the National Academy of Sciences, 111*(20), 7379–7384. https://doi.org/10.1073/pnas.1322174111

57. Flores-Kanter, P. E., Moretti, L., & Medrano, L. A. (2021). A narrative review of emotion regulation process in stress and recovery phases. *Heliyon, 7*(6), e07218. https://doi.org/10.1016/j.heliyon.2021.e07218

58. Hoge, E. A., Bui, E., Palitz, S. A., Schwarz, N. R., Owens, M. E., Johnston, J. M., Pollack, M. H., & Simon, N. M. (2018). The effect of mindfulness meditation training on biological acute stress responses in generalized anxiety disorder. *Psychiatry Research, 262*, 328–332. https://doi.org/10.1016/j.psychres.2017.01.006

59. Ryan, M., & Ryznar, R. (2022). The Molecular Basis of Resilience: A Narrative Review. *Frontiers in Psychiatry, 13.* https://www.frontiersin.org/journals/psychiatry/articles/10.3389/fpsyt.2022.856998

60. Danvers, A. F., Efinger, L. D., Mehl, M. R., Helm, P. J., Raison, C. L., Polsinelli, A. J., Moseley, S. A., & Sbarra, D. A. (2023). Loneliness and time alone in everyday life: A descriptive-exploratory study of subjective and objective social isolation. *Journal of Research in Personality, 107*, 104426. https://doi.org/10.1016/j.jrp.2023.104426

61. Mueller, C. M., & Dweck, C. S. (1998). Praise for intelligence can undermine children's motivation and performance. *Journal of Personality and Social Psychology, 75*(1), 33–52. https://doi.org/10.1037/0022-3514.75.1.33

62. Mangels, J. A., Butterfield, B., Lamb, J., Good, C., & Dweck, C. S. (2006). Why do beliefs about intelligence influence learning success? A social cognitive neuroscience model. *Social Cognitive and Affective Neuroscience, 1*(2), 75–86. https://doi.org/10.1093/scan/nsl013

63. Ehrlinger, J., Mitchum, A. L., & Dweck, C. S. (2016). Understanding overconfidence: Theories of intelligence, preferential attention, and distorted self-assessment. *Journal of Experimental Social Psychology, 63*, 94–100. https://doi.org/10.1016/j.jesp.2015.11.001

64. Gál, É., Tóth-Király, I., & Orosz, G. (2022). Fixed Intelligence Mindset, Self-Esteem, and Failure-Related Negative Emotions: A Cross-Cultural Mediation Model. *Frontiers in Psychology, 13*, 852638. https://doi.org/10.3389/fpsyg.2022.852638

65. Chen, L., Chang, H., Rudoler, J., Arnardottir, E., Zhang, Y., de los Angeles, C., & Menon, V. (2022). Cognitive training enhances growth mindset in children through plasticity of cortico-striatal circuits. *Npj Science of Learning, 7*(1), 1–10. https://doi.org/10.1038/s41539-022-00146-7

66. Koblinsky, N. D., Meusel, L.-A. C., Greenwood, C. E., & Anderson, N. D. (2021). Household physical activity is positively associated with gray matter volume in older adults. *BMC Geriatrics, 21*(1), 104. https://doi.org/10.1186/s12877-021-02054-8

References

67. Won, E., & Kim, Y.-K. (2020). Neuroinflammation-Associated Alterations of the Brain as Potential Neural Biomarkers in Anxiety Disorders. *International Journal of Molecular Sciences, 21*(18), 6546. https://doi.org/10.3390/ijms21186546

68. Dantzer, R. (2017). Role of the kynurenine metabolism pathway in inflammation-induced depression – Preclinical approaches. *Current Topics in Behavioral Neurosciences, 31,* 117–138. https://doi.org/10.1007/7854_2016_6

69. Mucher, P., Batmyagmar, D., Perkmann, T., Repl, M., Radakovics, A., Ponocny-Seliger, E., Lukas, I., Fritzer-Szekeres, M., Lehrner, J., Knogler, T., Tscholakoff, D., Fondi, M., Wagner, O. F., Winker, R., & Haslacher, H. (2021). Basal myokine levels are associated with quality of life and depressed mood in older adults. *Psychophysiology, 58*(5), e13799. https://doi.org/10.1111/psyp.13799

70. Lee, J. H., & Jun, H.-S. (2019). Role of Myokines in Regulating Skeletal Muscle Mass and Function. *Frontiers in Physiology, 10.* https://www.frontiersin.org/articles/10.3389/fphys.2019.00042

71. Lee, B., Shin, M., Park, Y., Won, S.-Y., & Cho, K. S. (2021). Physical Exercise-Induced Myokines in Neurodegenerative Diseases. *International Journal of Molecular Sciences, 22*(11), 5795. https://doi.org/10.3390/ijms22115795

72. Yang, T., Nie, Z., Shu, H., Kuang, Y., Chen, X., Cheng, J., Yu, S., & Liu, H. (2020). The Role of BDNF on Neural Plasticity in Depression. *Frontiers in Cellular Neuroscience, 14.* https://www.frontiersin.org/article/10.3389/fncel.2020.00082

73. Porter, G. A., & O'Connor, J. C. (2022). Brain-derived neurotrophic factor and inflammation in depression:

Pathogenic partners in crime? *World Journal of Psychiatry,* *12*(1), 77–97. https://doi.org/10.5498/wjp.v12.i1.77

74. Arosio, B., Guerini, F. R., Voshaar, R. C. O., & Aprahamian, I. (2021). Blood Brain-Derived Neurotrophic Factor (BDNF) and Major Depression: Do We Have a Translational Perspective? *Frontiers in Behavioral Neuroscience, 15.* https://www.frontiersin.org/articles/10.3389/fnbeh.2021.626906

75. Severinsen, M. C. K., & Pedersen, B. K. (2020). Muscle – Organ Crosstalk: The Emerging Roles of Myokines. *Endocrine Reviews, 41*(4), 594–609. https://doi.org/10.1210/endrev/bnaa016

76. Cassilhas, R. C., Antunes, H. K. M., Tufik, S., & de Mello, M. T. (2010). Mood, anxiety, and serum IGF-1 in elderly men given 24 weeks of high resistance exercise. *Perceptual and Motor Skills, 110*(1), 265–276. https://doi.org/10.2466/PMS.110.1.265-276

77. Maglio, L. E., Noriega-Prieto, J. A., Maroto, I. B., Martin-Cortecero, J., Muñoz-Callejas, A., Callejo-Móstoles, M., & Fernández de Sevilla, D. (2021). IGF-1 facilitates extinction of conditioned fear. *eLife, 10,* e67267. https://doi.org/10.7554/eLife.67267

78. De Voogd, L. D., & Phelps, E. A. (2020). A cognitively demanding working-memory intervention enhances extinction. *Scientific Reports, 10*(1), 7020. https://doi.org/10.1038/s41598-020-63811-0

79. Sujkowski, A. L., Hong, L., Wessells, R. J., & Todi, S. V. (2022). The Protective Role of Exercise Against Age-Related Neurodegeneration. *Ageing Research Reviews, 74,* 101543. https://doi.org/10.1016/j.arr.2021.101543

80. The Neuro Experience Podcast with Louisa Nicola

81. Hutchison, I. C., & Rathore, S. (2015). The role of REM sleep theta activity in emotional memory. *Frontiers in Psychology*, *6*. https://www.frontiersin.org/journals/psychology/articles/10.3389/fpsyg.2015.01439

82. Wittert, G. (2014). The relationship between sleep disorders and testosterone in men. *Asian Journal of Andrology*, *16*(2), 262–265. https://doi.org/10.4103/1008-682X.122586

83. Möller-Levet, C. S., Archer, S. N., Bucca, G., Laing, E. E., Slak, A., Kabiljo, R., Lo, J. C. Y., Santhi, N., von Schantz, M., Smith, C. P., & Dijk, D.-J. (2013). Effects of insufficient sleep on circadian rhythmicity and expression amplitude of the human blood transcriptome. *Proceedings of the National Academy of Sciences*, *110*(12), E1132–E1141. https://doi.org/10.1073/pnas.1217154110

84. Suri, R. (2002). TD Models of reward predictive responses in dopamine neurons. *Neural Networks : The Official Journal of the International Neural Network Society*, *15*, 523–533. https://doi.org/10.1016/S0893-6080(02)00046-1

85. Schultz, W. (2016). Dopamine reward prediction error coding. *Dialogues in Clinical Neuroscience*, *18*, 23–32. https://doi.org/10.31887/DCNS.2016.18.1/wschultz

86. Srámek, P., Simecková, M., Janský, L., Savlíková, J., & Vybíral, S. (2000). Human physiological responses to immersion into water of different temperatures. *European Journal of Applied Physiology*, *81*(5), 436–442. https://doi.org/10.1007/s004210050065

87. Ratan, Z. A., Parrish, A.-M., Alotaibi, M. S., & Hosseinzadeh, H. (2022). Prevalence of Smartphone Addiction and Its Association with Sociodemographic, Physical and Mental Well-Being: A Cross-Sectional Study among the Young Adults of Bangladesh. *International Journal of Environmental Research and Public Health*, *19*(24), 16583. https://doi.org/10.3390/ijerph192416583

88. Creswell, J. D., Bursley, J. K., & Satpute, A. B. (2013). Neural reactivation links unconscious thought to decision-making performance. *Social Cognitive and Affective Neuroscience*, *8*(8), 863–869. https://doi.org/10.1093/scan/nst004

Picture credits

Figure 1, p. 11: iStock.com/BulgakovaKristina

Figures 3 and 5: © 2015 Payne, Levine and Crane-Godreau (pp. 41 and 47)

Figure 4, p. 45: iStock.com/ngupakarti

Figure 6, p. 59: Credit: 'Plutchik's Wheel of Emotions', by Machine Elf 1735, available at https://en.wikipedia.org/wiki/Robert_Plutchik#/media/File:Plutchik-wheel.svg

Figure 7, p. 86 Credit: © "versusthemachines" and "The Decision Lab". Available from https://thedecisionlab.com/biases/negativity-bias.

Figure 9, p. 104: Credit: 'Salience Network' by Nekovaroca, Fajnerova, Horacek and Spaniel, available under a Creative Commons Attribution License 3.0 at https://commons.wikimedia.org/wiki/File:Fnbeh-08-00171-g002.jpg

Figure 13, p. 155: Credit: 'Maslow's Hierarchy of Needs' by Androidmarsexpress, available under a Creative Commons Attribution License 4.0 at https://en.wikipedia.org/wiki/Maslow%27s_hierarchy_of_needs#/media/File:Maslow's_Hierarchy_of_Needs2.svg

Picture credits

Acknowledgements

First and foremost, I am deeply grateful to my fiancé, Jorge, whose unwavering belief in me and constant encouragement propelled me forward even when I was in doubt. Thank you for holding space for me throughout my entire academic career and during the writing process of this book. You have been there since the beginning.

I am deeply grateful to the countless scientists and researchers whose groundbreaking work in neuroscience and psychology has illuminated the complex workings of the human mind. Their tireless dedication and collective efforts to pursue knowledge and share their findings with the world have provided the framework upon which this book is built.

Of course, I am indebted to my agent, Abigail Bergstrom, for being the biggest badass I could ever have asked for; I am ever so grateful to have found you. After I got off our initial call, I cancelled all meetings with other agents because I knew I only wanted to work with you. And to my editor, Karolina Kaim, for fighting to have me in her corner. As I said in the book, we are a match made in writing heaven. Cheers to many more.

I would like to thank my supervisor Dr Kristofer Kinsey for his support on this book, especially whilst completing my MSc research alongside it; my colleagues and friends, Dr Sula

Acknowledgements

Windgassen, Nas Fatih, Matthew Watson, Louisa Nicola, Rachelle Summers, Liadan Gunter, Isabel Taylor and Courtney Kremler who provided scientific insight, support and constructive criticism to my work.

I would also like to express my gratitude to my team at Penguin Michael Joseph in the UK and my team at HarperOne in the US whose professionalism and dedication helped bring this book to fruition. I'd also like to thank my translation agent, Alexandra Cliff, who has brought this book into more territories across the world.

This wouldn't be an acknowledgement section if I didn't mention my mother, Sheila Van Der Merwe, who gave birth to a legend, because she is one herself. Thank you for loving me so much throughout our turbulent childhood. It is through love and human connection that we heal . . . Love invents us. I'd also like to extend my gratitude to my mother-in-law and clinical psychologist, Kim Camacho, for heavily shaping my integrity as a scientist; but more importantly, for being a shoulder to lean on when I'm in the depths of despair. It is a blessing to get a second motherly figure in one's life when they are like you. Also, thanks for birthing your son. Thanks to my father-in-law, Luis Camacho, for letting, Jorge, the dogs and I, live in his beautiful home during the writing process, and wishing we could live together as a family forever. Sorry, that sounds cute but practically speaking we would eventually need our own space.

To my dear siblings, Silvy and Dylan Vignola, thank you for existing. Love lifts us up where we belong, and we are meant to be related. I love you. To my closest friends, Lucia Rippa,

Gabriella Baleta, Ubaka Onyekwelu, Caroline Guiterrez, Louisa Russel-Henry, Shannon Botha and Jo Hargreaves for always bringing the vibes and banter, filling my life with joy, and for always answering the phone when I need you.

Last but not least, to the two beings I love the most in this whole world (sorry to everyone else), Kobe and Max, my two dogs. You don't know how to read but I'll come and give you a big smooch in a minute to show you how grateful I am that we found each other.

From the bottom of my heart, I'd like to thank each and every individual who played a part, big or small, in bringing this book to life. If I haven't mentioned you, because I have likely forgotten, but if you feel personally insulted by this, please pick up the phone and let's talk it out.